Vanilla, Chocolate, & Strawberry

Vanilla, Chocolate, & Strawberry

The Story of Your Favorite Flavors

Bonnie Busenberg

Lerner Publications Company ▪ Minneapolis

Note: Words in **boldface** type are defined
in a glossary that begins on page 106.

Library of Congress Cataloging-in-Publication Data

Busenberg, Bonnie.
    Vanilla, chocolate, and strawberry : the story of your favorite
flavors / Bonnie Busenberg.
        p. cm.
    Includes index.
    Summary: Describes how vanilla, chocolate, and strawberry came to
become popular flavorings, how they were originally used, how
they're used today, and what makes them taste the way they do.
Includes recipes.
    ISBN 0-8225-1573-3
    1. Flavoring essences—Juvenile literature. 2. Vanilla—juvenile
literature. 3. Chocolate—Juvenile literature. 4. Strawberries—
Juvenile literature. [1. Flavoring essences. 2. Vanilla.
3. Chocolate. 4. Strawberries.] I. Title.
TP418.B87 1994
664'.5—dc20   93-15101
                                                         CIP
                                                          AC

Manufactured in the United States of America
1 2 3 4 5 6 – I/JR – 99 98 97 96 95 94

For George, John, and Stavros:
different flavors all, each totally delectable

Contents

Cacao beans are spread out in the sun to dry. After they are dried and roasted, the beans will be processed into chocolate.

Introduction

On a summer evening, when you and your family walk into an ice-cream shop, you're faced with dozens of luscious flavors: everything from bubble gum to passion fruit, from mocha fudge to blueberry cheesecake. The choices seem limitless. There are three flavors, though, that are sure to be on the menu. They are old standbys, the all-time favorites that people come back to again and again, no matter what the current food fads are.

Vanilla, chocolate, and strawberry—everyone recognizes these flavors. They are very different from one another, yet they have a lot in common. They have been enjoyed by some societies for more than 500 years, but they exploded in popularity during the 18th and 19th centuries. At one time, each of these flavors was mistakenly thought to be an important medicine.

Vanilla, chocolate, and strawberry are such common tastes that few people give them much thought. Do you know where these flavors come from? Do you know how they got their names, how they're prepared, and how they were originally used? What makes them taste the way they do, and how have they changed and been

table sugar
$(C_{12}H_{22}O_{11})$

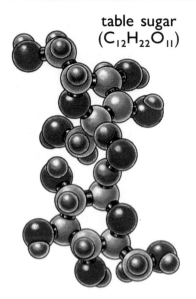

Chemical Compounds

Chemical compounds are made up of basic elements combined in particular proportions. Salt, for example, is a compound made up of one atom of sodium (written as *Na* by chemists) and one atom of chlorine (Cl). Chemists write salt as NaCl. The white sugar we eat, called sucrose by chemists and written $C_{12}H_{22}O_{11}$, is a compound made up of 12 atoms of carbon (C), 22 atoms of hydrogen (H), and 11 atoms of oxygen (O). Chemical compounds may have properties that are very different from the elements they are made of. For example, water (H_2O) is a liquid, although it is made from hydrogen (H) and oxygen (O), which are both gases.

imitated over the years? Most people don't know the answers to these questions. Flavor is something we tend to take for granted. We're glad it's in the food we eat, but we don't worry about what causes it or how we recognize it.

What Is Flavor?

Many people would say that flavor is the way something tastes. That definition is only partly correct. When you have a head cold, have you noticed how "flat" food tastes? Have you ever held your nose in order to swallow a nasty-tasting medicine? If so, then you realize that the sense of smell has a lot to do with what we usually think of as taste. In fact, flavor is the combined sensation of a food's taste and odor, as they are perceived by the eater.

Humans are much more sensitive to smell than they are to taste. While we can perceive only four basic categories of taste—sweet, salty, sour, and bitter—we recognize and remember thousands of different odors. What's more, we can detect aromas even when they are present in very small amounts. Most experts agree that smell contributes more than taste does to our perception of flavor. The texture and temperature of a food also significantly affect its flavor.

Humans perceive flavor by means of special nerve cells called **receptors,** which transmit nerve impulses, or messages, to the brain. The receptor cells that are responsible for taste are known as taste buds or papillae. They are located mostly on the tongue. The receptor cells responsible for smell, called olfactory receptors, are located in the lining of the nose and the back of the nasal cavity. When you eat, the taste buds on your tongue are stimulated by chemical compounds in the food.

As food is chewed, it gives off vapors and molecules

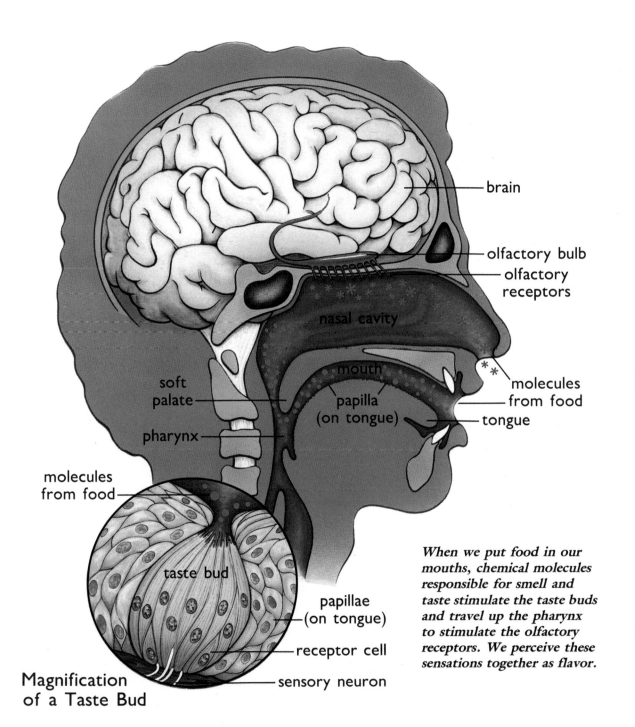

brain

olfactory bulb

olfactory receptors

nasal cavity

mouth

molecules from food

tongue

soft palate

papilla (on tongue)

pharynx

molecules from food

taste bud

papillae (on tongue)

receptor cell

sensory neuron

Magnification of a Taste Bud

When we put food in our mouths, chemical molecules responsible for smell and taste stimulate the taste buds and travel up the pharynx to stimulate the olfactory receptors. We perceive these sensations together as flavor.

Taste Buds

Taste buds do not all have the same shape. If you look at them under a microscope, some look like tiny mushrooms, others like miniature hills surrounded by moats, and still others like tiny threads or cones. At one time, it was thought that each type of taste bud might be a specific receptor for one of the four basic taste sensations (sour, sweet, bitter, and salty). Now it has been shown that all taste receptor cells are sensitive to three or four of the basic tastes. At the same time, though, certain areas of the tongue are particularly sensitive to one or more of the taste sensations. Sweetness is perceived most readily on the tip of the tongue, bitterness at the back, while salty and sour tastes are perceived on the sides.

of gases in the mouth. These gases pass through the pharynx, a flattened tube leading from the back of the mouth into the nasal cavity. There the gases stimulate the olfactory receptors.

Humans have thousands of taste buds and olfactory receptors in their noses and mouths. These specialized cells are stimulated when we smell and eat food, and they respond by sending a complex pattern of nerve impulses to the brain. The receptor cells are not all alike. When food is placed in the mouth, some cells might increase the frequency of nerve impulses they send to the brain, while others decrease them. The brain translates this coded pattern of impulses received from all the different receptor cells into a perception of flavor.

No one is sure how receptor cells interact with chemicals in foods. While many theories are being investigated, common phenomena remain unexplained. For example, why do we think saccharin and sugar taste similar? Saccharin molecules are different in shape, size, and chemical composition from sugar molecules such as glucose or sucrose. And why do our smell receptors become blocked—unable to transmit signals to the brain—when they are exposed to a powerful odor for a long time? (Have you noticed how you can "get used to" a bad smell after a while?)

Sources of Food Flavorings

Many foods have a strong natural flavor. Flavorings are added to other foods to improve their taste and aroma. Most natural flavors come from some part of a plant: roots, stems, leaves, flowers, or fruits. The chemicals in the plant that produce the flavor might be found throughout the plant or only in certain parts. They may be present when the plant is green (unripe) or only when it is ripe. In some plants, flavor chemicals

develop when part of the plant is heated, **fermented,** or treated in some other way.

Herbs provide natural flavorings for food.

Humans have added fresh or dried plant parts to foods for thousands of years. The cultivation of plant crops for use as flavorings represents an important segment of world agriculture, and the processing and sale of natural flavoring materials is a major global industry.

Natural flavors derived from plants have always been rather costly and limited in supply, however. Over the past 100 years, scientists have discovered ways to isolate and **concentrate** the flavor-causing chemicals from plants. These concentrates (also called extracts, distillates,

Some flavors, such as hot pepper, come from plants (above), while other flavors, like bubble gum, are created artificially.

tinctures, and resins) can be stored and handled more easily than the plant itself.

More recently, modern technology has allowed researchers to identify many of the chemicals responsible for flavor. When the chemical makeup of a flavor is known, it can often be created in a laboratory. Chemists are able to make many artificial flavor compounds that resemble natural flavors. Flavor chemists have also created completely new flavors that have no precise natural counterpart—flavors such as bubble gum, spice, and smoky. Synthetic or artificial flavors are much less expensive to produce than natural flavors, so they are widely used in commercial products.

Currently, more than 200 natural flavors and 750 artificial flavorings are used in the food industry. Each flavor has a story all its own—about the plant origins of the flavor, the way the plant was used through the centuries, the attempts to recreate the flavor in a chemistry laboratory, and the men and women who were involved in all these processes.

This book tells three of these stories. You already know that vanilla, chocolate, and strawberry are valued for their tastes and aromas. But you are about to discover that the history, biology, and chemistry of these three substances are as rich, complex, and enjoyable as the flavors themselves. Read on if you have an appetite for myth, mystery, innovation, and scientific adventure!

Chapter 1

Vanilla:
The Little Pod

Pills and other pharmaceutical products often contain vanilla.

When you hear the word *vanilla,* what do you think of? Ice cream? Pudding? Milk shakes? Something delicious, of course. Vanilla is one of the most widely used flavors in sweets of all kinds, including candy, soft drinks, and baked goods. But did you know that vanilla is often used to mellow the flavor and smell of tobacco? It's the major flavor in several liqueurs, and it's frequently used in the making of cough syrups, pills, and other medicines.

What's more, vanilla is practically inseparable from chocolate! Strange as it may seem, vanilla is almost always used in the making of chocolate products, such as chocolate candy bars, chocolate syrup, and hot chocolate drinks. If you look in a cookbook, you will see that recipes for brownies, fudge, chocolate cakes, chocolate frostings, and chocolate chip cookies usually call for the addition of vanilla. The two flavors were first used together more than 500 years ago, in drinks prepared by the Indians who lived in the tropical areas of Mexico. Europeans were introduced to both chocolate and vanilla through the hospitality of the great Aztec emperor, Montezuma II.

The Discovery of Vanilla

The vanilla plant grows naturally in the rain forests of the Caribbean and Central America, on the seacoast of Mexico, and in the northernmost regions of South America. Since vanilla could be found only in the tropical regions of the Americas, it was unknown in Europe until the end of the 1500s.

In 1519, when the Spanish conqueror Hernán Cortés and his conquistadores were initially welcomed into the flourishing Aztec civilization in Mexico, they observed that the emperor Montezuma II had a passion for a drink called **chocolatl.** The concoction was made from powdered **cacao** beans and ground corn. It was flavored with honey and **tlilxochitl** (vanilla), the ground seed-pod of an orchid plant. The dried and shriveled seed-pod was about the size of a string bean, dark brown in color, and waxy on the outside. It had a rich, mellow aroma.

The Aztecs obtained their tlilxochitl from another people, the Totonacs, who lived in the Gulf coast region

An Aztec drawing records the meeting of Montezuma and Cortés.

of Mexico in what is now the state of Veracruz. Tlilxo-chitl and the plant it came from, called "xanath," played an important role in Totonac culture. The people used it as a perfume, a flavoring for food and drinks, a medicine, a love potion or aphrodisiac, and an insect repellent. According to Totonac mythology, the plant was a gift from the goddess who bore the same name.

The Totonacs believed that when the world was young and fresh, a beautiful young goddess named Xanath, the daughter of the powerful goddess of fertility, came to earth and fell in love with a Totonac warrior. The goddess could not transform herself into a mortal woman, nor could she make her lover immortal. Xanath could neither marry the warrior nor abandon him. To solve her dilemma, the goddess changed herself into a vanilla vine and bestowed herself upon her lover and his people. That way, with the flowering of the sacred plant, she would provide the people with an eternal source of happiness.

The Totonac people discovered the delicious secret hidden in the seedpod of the vanilla orchid plant.

The Aztecs conquered the Totonacs around the year 1425. The tribute the Aztecs demanded from the conquered Totonacs was a portion of their annual tlilxochitl harvest. The Totonacs had learned the secrets of how to cultivate the vanilla plant and process its fruit. They, and after them the Aztecs, used the plant to flavor their chocolatl.

How the Totonacs learned to cultivate the vanilla plant and ensure good harvests of the fruit is a mystery lost in the past. But if these Native Americans had not discovered how to prepare the vanilla seedpod so that it would release its hidden flavor, Europeans might never have paid any attention to the plant. European explorers probably expected to find new spices in the tropical areas of the Americas. After all, many of the spices and flavorings that had been used for centuries in Europe came from the tropics of Africa and Asia. But the explorers would not have expected to discover a new flavor in the fruit of an orchid plant. Even though there are many types of orchids—more than 35,000 different species— vanilla is the only flavor ever derived from any of them.

What's more, the growing plant gives little hint of the potential treasure hidden inside the fruit. The flower has some fragrance, but the unripe seedpod has neither the taste nor the smell of vanilla. Only after the seedpod is picked unripe and **cured** for several months does it develop the properties that make it such a wonderful addition to food.

Introducing a New Flavor

The Spanish soldiers renamed the fragrant tlilxochitl pod *vainilla,* which means "little pod" in Spanish. (The word became "vanilla" in English.) Cortés took vanilla pods back to Spain with him to use in the preparation of chocolatl, which quickly became a favorite drink at the royal court.

Amazingly, almost 80 years passed before someone realized that vanilla was a delicious flavor in its own right and could be used in foods other than the chocolate drink. Hugh Morgan, who prepared herbal medicines for Queen Elizabeth I of England, made this discovery in 1602. The queen developed quite a liking for vanilla. It is said that late in her life she ate only foods that were flavored with vanilla.

The French, too, developed an intense fondness for vanilla. In the 1700s, the flavor was used more in France than in any other European country. Thomas Jefferson learned about vanilla in Paris while he was serving as U.S. ambassador to France. Jefferson brought vanilla pods back to America with him and introduced the wonderful new flavor to the young country. The use of vanilla became so common that Americans eventually began

Queen Elizabeth I adored the taste of vanilla.

to use the word *vanilla* to mean plain or ordinary—something it definitely is not.

The vanilla orchid flower is light lemon yellow in color and is smaller than the plant's leaves.

The Vanilla Orchid Plant

The scientific name of the vanilla orchid is *Vanilla planifolia*. The plant is a vine that thrives in hot, humid tropical environments. (Tropical regions lie between 20° north and 20° south of the equator.) The natural habitats for the vanilla plant are areas below 2,500 feet in altitude that receive about 100 inches of rain per year. It grows best in areas where 10 months of rainfall are followed by a 2-month dry season. The plant prefers the rich soil and good drainage found on gently sloping land.

The wild vanilla vine, with its broad, flat leaves, uses forest trees as support as it grows. It often climbs 50 feet or more to the top of tall trees. Although the plant has a root system below the ground, it also produces slender rootlets from the leaf nodes on the stem. These rootlets not only absorb water from the humid air but also help attach the vine to a tree by twining around the tree's smaller branches. The vanilla plant is not a parasite, however. It does not steal nutrients from the tree it grows on. The vine uses the tree for support only, obtaining all its nutrients from the soil and the air.

While many orchid flowers are strikingly beautiful in shape and color, the vanilla orchid is rather plain. The pale greenish yellow flower with a satiny texture has a curious arrangement of parts, though. It appears to have five elongated petals surrounding a tube. In fact, the flower has three **sepals** and three petals. In many plants, the sepals, located just beneath the petals, are green and look like small leaves. They function to support and protect the delicate petals.

In the vanilla orchid, as in most orchids, the sepals are large and practically indistinguishable from two of the petals. The third petal is unique, however. It is larger and has a fringed lip that thickens in the middle. This petal curves upward, forming a sort of tube that surrounds the flower's reproductive parts. The male and female reproductive parts of the flower, called the **stamen** and **pistil,** also have an unusual arrangement in the vanilla orchid. They are fused together, forming a central column in the flower. The column is somewhat hidden by the curving petal.

Many plants reproduce sexually, as animals do, through the union of male and female reproductive (sex) cells. The union results in a new plant or animal that is similar to but not exactly like its parents. In flowers, **sexual reproduction** results in the creation of seeds. For sexual reproduction to occur, the male sex cells contained in the yellow, powdery **pollen** of the stamens must reach the female sex cells, located in the **ovules** of the pistil.

The all-important transfer of pollen from stamen to pistil, or **pollination,** occurs in different ways in different

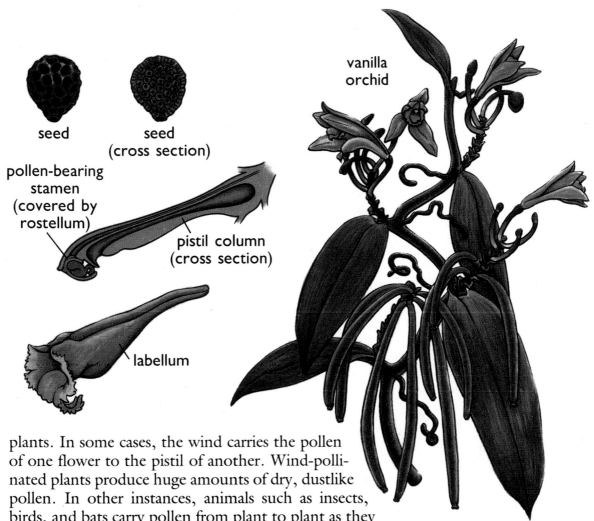

seed

seed
(cross section)

pollen-bearing
stamen
(covered by
rostellum)

pistil column
(cross section)

labellum

vanilla
orchid

plants. In some cases, the wind carries the pollen of one flower to the pistil of another. Wind-pollinated plants produce huge amounts of dry, dustlike pollen. In other instances, animals such as insects, birds, and bats carry pollen from plant to plant as they visit flowers in search of food. Animal-pollinated flowers usually have nectar, aroma, or colored petals to attract their pollen-carrying visitors. Some, but not many, flowers can pollinate themselves. Most orchids, including the vanilla orchid, are animal-pollinated.

When a pollen grain lands on the sticky surface of the pistil (called the stigma), the pollen forms a tube down into the ovary, which contains the ovules. Sperm cells (male sex cells) travel down the tube, and when one of the sperm cells joins with an egg (female sex cell) in the ovule, fertilization occurs. Then the ovule begins to develop into a seed.

The vanilla plant, including flowers and green seedpods. The large fringed petal of the flower is called the labellum. It curves upward to surround the reproductive column. The flower's pollen, located at the top of the column, is very sticky. It is covered by a flap of tissue called the rostellum.

Pollination of the Vanilla Flower

Because of the structure of the flower, orchids are difficult to pollinate. The pollen, which is covered by a flap (the rostellum), must reach the stigma of the flower, located farther down the column. The flap and the stickiness of the pollen prevent the wind from carrying the pollen from one flower to another. Pollination is often carried out by bees. The insect enters the fringed petal (labellum) of the orchid to feed on nectar beneath the column. In backing out of the flower, the insect often knocks off the rostellum from the top of the column, exposing the pollen. As the bee continues to back out of the flower, the pollen sticks to its back. When the bee visits another flower to feed, in backing out of the second flower, the pollen carried on the bee's back is brushed onto the sticky stigmatic surface of the column. From there, the pollen travels down to the ovary and pollinates the flower. The bee, continuing to back out of the second flower, will then knock off the rostellum cap from the second flower and pick up pollen, which it will bring to the next flower it visits.

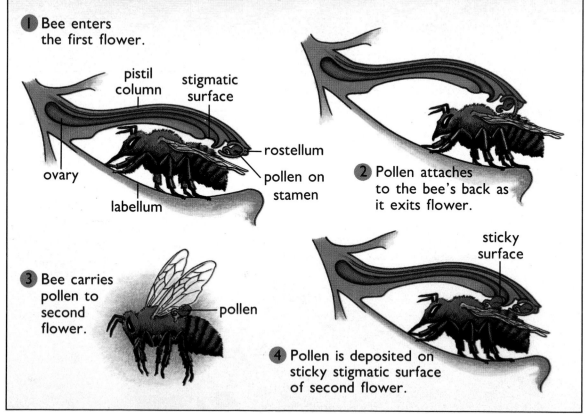

1 Bee enters the first flower.

pistil column — stigmatic surface

rostellum

pollen on stamen

ovary

labellum

2 Pollen attaches to the bee's back as it exits flower.

sticky surface

3 Bee carries pollen to second flower.

pollen

4 Pollen is deposited on sticky stigmatic surface of second flower.

After the vanilla flower is pollinated, usually by bees or small birds, in four to nine months it produces an elongated pod containing millions of tiny seeds. Although vanilla seeds are much larger than most orchid seeds, they are still very small—about the size of ground black pepper grains.

The vanilla plant grows easily in the tropics of the Americas. First the Totonacs and then the Maya and Aztec people cultivated the vanilla plant in plantation-like areas. The Indians trimmed the trees in part of the jungle and tied a 4-foot piece of vanilla vine to the trimmed tree trunks. Soon the vines sent out roots into the soil. Using the tree as support, the vines grew, flowered, and produced fruit. The Indians learned that if the vine was allowed to grow directly upward, it would not flower well, so they looped each vine back to the ground once it reached 5 feet high.

The slender green vanilla pods are about 5 to 10 inches long.

When the Spanish army invaded and conquered the Aztec nation, the soldiers copied Aztec techniques and established their own vanilla plantations in Mexico. Later, Spanish and French colonists tried to grow the plants in other tropical areas of the world. But a curse seemed to hang over their efforts. Try as they might, for 300 years no one could get the vine to produce fruit when it was planted outside the Americas. In places like Java and Madagascar, which have climates and conditions similar to those of tropical Mexico, the vanilla plants grew very well and produced abundant flowers. But to the frustration of European gardeners in the tropics, the vines never produced the long fruit pods.

The problem baffled plantation owners until 1836, when a Belgian **botanist,** Charles Morren, went to the rain forests of Mexico to study the vanilla orchid. He noticed that the male and female parts of the vanilla orchid are separated by a flap of tissue, the rostellum, which prevents the flower from pollinating itself. In

Hand-Pollination of the Vanilla Orchid Flower

Holding an individual flower, the human pollinator uses the pointed tip of a small bamboo stick to pry open the rostellum and pick up the sticky male pollen bundles. Then the worker smears the pollen mass onto the stigma (the female organ, located in a shallow hole on the column) with the left thumb.

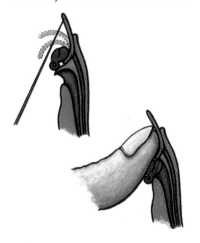

Mexico and Central America, Morren observed that the plants were usually pollinated by melipone bees, which have very long mouth parts, or by certain species of hummingbirds. Morren guessed that insects and birds outside of the Americas were not adapted to effectively pollinate these curious flowers. The plantation owners had transplanted the orchids—but not the pollinators. Without pollination, vanilla pods could not be produced. To solve this problem, Morren suggested that plantation owners hand-pollinate the flowers.

Birds and bees naturally (and unknowingly) collect pollen on their bodies as they feed on one flower and carry the pollen to the next flower they visit. In hand-pollination, humans take the place of these natural pollinators. Workers must carefully collect pollen and transfer it to the stigmas (female receptive surfaces) of the flowers. This is a tedious business. For vanilla orchids, hand-pollination is particularly difficult because the flowers bloom for one day only, and they must be pollinated in the early hours of that day for fertilization and fruit development to occur.

In 1841, Edmond Albius, a former slave from the French-governed island of Réunion, perfected an effective, quick hand-pollination method that used the pointed tip of a small bamboo stick. The technique was simple, but it was the crucial innovation that allowed vanilla plantations to be established in Réunion, Java, Mauritius, Madagascar, Tahiti, the Seychelles Islands, the Philippines, the Comoro Islands, Tanzania, Jamaica, and other islands in the West Indies. Even Mexican plantations prefer to use hand-pollination today, despite the fact that natural pollinators are available to do the job. Ironically, hundreds of years ago, the Totonacs had also hand-pollinated their vanilla plants to ensure a high rate of pod production. They called the procedure the "marriage of vanilla."

Growing Vanilla Orchids

When growers cultivate vanilla plants today, it can take up to five years before any of the precious fruit is ready to be harvested. On a vanilla plantation, fast-growing trees and shrubs are planted a year before the vanilla vine is planted. These trees and shrubs, planted about 8 feet apart, act as living supports for the vanilla plants. The best live supports are evergreen trees with small leaves, nonpeeling bark, and low branches. A year later, workers plant cuttings from mature vanilla vines close to the base of the young support trees, attaching them to the trees. Cuttings (1- to 4-foot sections of the vanilla plant) are used instead of seeds because orchids are notoriously difficult to grow from seeds.

It takes the young cuttings 10 to 12 weeks to establish themselves and sprout new roots. The young vines grow vigorously and must be pruned (cut back) constantly so that they will not grow beyond the reach of the workers who must hand-pollinate the flowers and harvest the pods. On plantations, vanilla vines are harvested for 10 to 12 seasons before they are replaced by new, more productive plants.

Vanilla orchids grow in clusters of 20 or more, suspended on short stalks that sprout from the stem of the vine. A healthy vine may produce as many as 1,000 flowers every year. Only 40 to 50 large, perfect blossoms are selected from each vine to be hand-pollinated, however, to ensure that all the resulting pods will receive enough nourishment. No more than six to eight flowers are chosen from any one cluster so that the pods will not be crowded and stunted in their growth. The flowering season in a plantation may extend over two months, but each flower lasts just one day. Workers must be quick and efficient. A skilled worker can hand-pollinate over 1,000 blossoms a day.

The stem of the vanilla vine is fleshy and succulent. The leaves are bright green, smooth, and oval shaped. The veins in the leaves run parallel to one another, as they do in grasses, irises, and lilies. Twining roots grow upward from the stem, allowing the vine to cling to its tree support and collect water.

After the vanilla flowers are pollinated, it takes another four to nine months for the seedpods to mature. They are picked before they become fully ripe, just as their color starts to change from green to yellow. When the pods are picked from the vine, they are about the size of a marker pen, 5 to 10 inches long and 1 inch wide.

The annual harvest of the fruits of the vanilla orchid is not the end of the story. It is just the beginning of another process that must be carefully carried out to develop the flavor in the vanilla pods.

Curing Vanilla Pods

After the vanilla crop is harvested, the grower must cure, or preserve, the pods. In the curing process, the vanilla pods undergo controlled fermentation. During fermentation, enzymes—compounds that control chemical reactions within cells—cause complex molecules to break apart into smaller components. A chemical compound known as **vanillin,** part of a large molecule in the fresh vanilla pod, is released during fermentation. Vanillin gives the cured pod much of its flavor and aroma, which the fresh pod lacks.

Curing is a long and delicate process, but it is not unique to the vanilla pod. Other products, including cacao beans and coffee beans, also must be cured before they are marketed.

Like wine makers, different vanilla growers have their own special techniques for curing the pods. But all growers follow the same basic steps. First, the pods are heated to kill the living seed tissue. Then they are subjected to "sweating," which starts the fermentation. Next follows a period of drying, during which the fermentation slows down and eventually ends. Finally, the pods are aged in metal-lined boxes to mellow the flavor and aroma.

The first step in making vanilla is to cure the green pods.

The initial heating of the vanilla pods is traditionally done by laying them in the sun. Makers of Bourbon vanilla (vanilla that comes from the islands of Madagascar, the Comoro Islands, and Réunion, which was formerly known as Bourbon) immerse the fresh beans in hot water instead. To sweat the vanilla pods, workers spread them out in the sun for several hours, then fold them in blankets or boxes to "sweat" until the following morning. This sweating process may be repeated for 10 to 20 days, until the pods are supple and deep brown in color. Pods are dried on trays in the sun or in heated rooms for many more weeks. Finally, the pods are placed in boxes to age for another two or three months, allowing the flavor to develop and mature.

The curing process can take as long as six months. After the vanilla pods are cured, they are sorted and graded according to quality, length, aroma, color, flexibility, and luster, or how glossy they are. The long, slim, pencil-like pods are then tied in bundles and packed in

After "sweating," the pods are spread out to dry in the sun for several weeks.

Workers sort cured vanilla pods, now called beans. The shiny dark brown beans are ready for market.

airtight tins to conserve their aroma. Prepared and stored in this manner, vanilla beans will keep for many years (cured vanilla pods are often called beans). Highest grade beans are sold to bakers and gourmet cooks, especially in Europe. Blends of various grades of vanilla are used for production of extracts and in the food industry.

During the curing process, vanilla beans lose almost 80 percent of their weight (mostly in the form of water). It takes 5 pounds of fresh pods to yield 1 pound of dried beans ready for the market. A grower can expect to harvest 600 pounds of fresh pods from each acre of a plantation. This will yield 120 pounds of cured, dried pods to be shipped to buyers.

The cured beans will be packed in tins and shipped around the world.

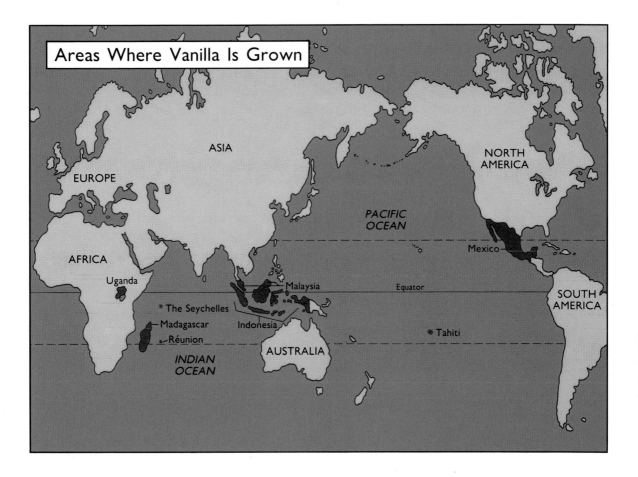

Areas Where Vanilla Is Grown

Several million pounds of vanilla are traded world-wide each year. The United States alone imports about three million pounds annually from the major vanilla-producing countries, which include Madagascar, Ré-union, the Seychelles Islands, Mexico, Tahiti, Indonesia, Malaysia, and Uganda.

The majority of vanilla plantations are located in Uganda, Madagascar, Réunion, the Seychelles Islands, Malaysia, Indonesia, Tahiti, and Mexico. Notice how close all these countries are to the equator.

The Chemical Vanillin

During the curing process, the chemical vanillin is released from the sugar molecules in the fresh pod and forms crystals on the outside of the pod. People soon

The Chemical Structure of Vanillin

In chemical terms, vanillin can be called 4-hydroxy-3-methoxy-benzaldehyde. Chemists draw the compound to look like this:

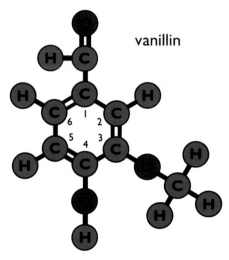

vanillin

The basic structure of the molecule is known as a benzene ring (six carbon molecules that bond together into a ring structure). In vanillin, carbons 2, 5, and 6 have a molecule of hydrogen attached to them. But in the remaining three carbons, the hydrogens have been replaced with more complicated groups. Carbon 1 has an aldehyde group (CHO) attached to it, carbon 3 has a methoxy group (H_3CO) attached to it, and carbon 4 has a hydroxyl group (HO) attached to it.

realized that this colorless crystalline substance was responsible for the wonderful taste and smell of the vanilla bean. By 1858 scientists were able to separate and remove vanillin from the cured pods.

Once chemists knew the chemical composition and structure of vanillin, they were soon able to make it in the laboratory. By 1874 chemists had succeeded in turning chemicals commonly found in pine tree sap into vanillin. By 1891 chemists had learned how to make vanillin from eugenol, a fragrant substance obtained from cloves. At about the same time, scientists also devised a way to produce vanillin from sugar. Now artificial vanilla is usually produced from a substance called lignin, a woody fiber found in wood and paper pulp.

The chemical vanillin is sometimes called synthetic vanilla or artificial vanilla, because it comes from a source other than true vanilla beans. Synthetic vanilla is much less expensive than real vanilla, and for that reason it is used widely in commercial food preparation. Unfortunately, artificial vanilla has a harsh smell and a bitter aftertaste. Real vanilla is not chemically pure. Its minor flavor components and secondary scents give it a delicate, spicy flavor and a complex aroma that are impossible to duplicate in the laboratory.

Vanilla in the Kitchen

If you have vanilla in your kitchen at home, it is probably a liquid in a small dark bottle, not a shriveled brown pod. In Europe many people prefer to cook with vanilla beans, but in the United States only professional chefs and dedicated cooks buy and use vanilla beans. Still, you can probably find one in the gourmet foods or spice section of your grocery store.

The black flecks that you see in some brands of "real vanilla" or "French vanilla" ice cream are the seeds of the vanilla bean. Pastry chefs often put a whole vanilla bean into their puddings or sauces. The bean's flavor is remarkably long-lasting, so the bean can be removed after being used in cooking, then washed, dried, and used many more times. Since vanilla is often an ingredient in sweets, many cooks put a vanilla bean into a container of granulated sugar so that the sugar will already have the flavor and smell of vanilla. The bean can be used over and over to flavor batches of sugar until it loses all its aroma. You can make vanilla sugar by putting two cups of granulated sugar into a sealed jar with a vanilla bean (bend the bean if it is too long). Seal the jar and let it sit for two weeks. Afterward, use the sugar to sweeten whipped cream and taste the difference.

In most American homes, vanilla extract is used for adding flavor to food. The extract is made at a factory, in a process that is very similar to percolating coffee. First the cured beans are chopped and placed in a large, stainless-steel basket. The basket is then placed in a warm alcohol-and-water solution inside an extraction vat. The extracting solution, which is about 35 percent alcohol, draws out the flavor of the beans as it percolates down through them to the bottom of the vat. From there it is recirculated to the top to continue percolating until as much vanilla flavor has been extracted from the beans as possible. When the flavor is strong enough, the liquid or extract is drawn off, strained, and aged for about a month to "polish" the flavor.

You can buy either pure vanilla extract or imitation vanilla extract. The pure extract ingredients are vanilla, alcohol, and sometimes sugar. The imitation vanilla extract ingredients are listed as water, alcohol, extracts of **cocoa,** tea, and other natural flavorings, vanillin and other artificial flavorings, and caramel color. You can detect the difference between the two products just by smelling the liquids in the bottles. The best way to

Gourmet cooks like to use real vanilla beans. If you leave a vanilla bean in a jar of sugar for two weeks, you can use the sugar in whipped cream and taste the vanilla flavor.

compare the two flavorings, though, is to put a few drops of each on separate sugar cubes and taste the cubes. Try the test on your family and friends and see if they can tell the difference between the two extracts. Have them guess which one is which. The real vanilla is stronger and deeper in both flavor and aroma. But some people who are used to eating synthetic vanilla may actually prefer the imitation extract.

Uses of Vanilla

The Totonac Indians used tlilxochitl as a medicine. Early European physicians also thought vanilla had medicinal properties. Some claimed it was a stimulant that would keep you awake, and others said it was good for the stomach and an antidote to poison. Romantics claimed it was a love potion. In America, vanilla was listed in the official U.S. *Pharmacopeia,* an encyclopedia of medicines from natural sources, from 1860 to 1910.

None of the claims about vanilla's beneficial medical effects has ever been scientifically proven. Doctors have shown that vanillin stimulates the liver to increase its production of bile, a green fluid that helps you digest your food. But vanillin has practically no effect when taken in the amounts usually found in foods.

On the other hand, both vanillin and vanilla beans sometimes affect humans adversely. The chemical, in both its natural and synthetic forms, can cause headaches. Some people experience itching, burning, stinging, reddening, or blistering if their skin has prolonged contact with vanilla. This condition is known as vanillism and usually only affects workers who sort and process vanilla beans.

Though useless as a medicine, vanilla is used in other products besides food. The wonderful scent of vanilla

Most people use imitation or pure vanilla extract in their cooking.

is sometimes used as a component of perfumes, particularly sweet floral perfumes. You can make your own simple perfume by sprinkling a few drops of vanilla extract on absorbent cotton and tucking it into your dresser drawers. Remember not to put it directly on your skin, because it may be irritating.

Do not be tempted to taste pure vanilla extract from a spoon. It is a strong concoction that at the very least will make your mouth pucker. Instead, to savor vanilla in its purest forms, you can try the recipes that follow.

People who handle vanilla beans all day run the risk of getting vanillism.

Vanilla Pudding

RECIPE FOR: Vanilla Pudding

INGREDIENTS:
2 cups milk, whole, lowfat, or nonfat
2½ tablespoons cornstarch
4 tablespoons sugar
⅛ teaspoon salt
1 teaspoon vanilla extract

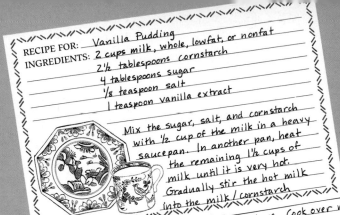

Mix the sugar, salt, and cornstarch with ½ cup of the milk in a heavy saucepan. In another pan, heat the remaining 1½ cups of milk until it is very hot. Gradually stir the hot milk into the milk/cornstarch mixture. Cook over moderately low heat, stirring constantly, until the mixture thickens (this happens quickly). Continue to cook the thickened mixture for another 10 minutes, stirring constantly so that the pudding does not stick to the bottom of the pan and burn. Remove the pudding from the heat, allow it to cool for 5 minutes, and stir in the vanilla extract. Spoon the pudding into two or three serving dishes. This dish may be served warm or refrigerated for several hours before serving. It is particularly good topped with fresh fruit slices.

SERVINGS: 2-3

Fluffy Vanilla Icing

RECIPE FOR: Fluffy Vanilla Icing

INGREDIENTS:
2 egg whites
1½ cups sugar
5 tablespoons ice water
¼ teaspoon cream of tartar
1½ teaspoons light corn syrup
1 teaspoon vanilla extract

Combine all ingredients, except vanilla, in the top of a quart-size double boiler and mix, using an electric hand mixer set at low speed. Bring water to a boil in the bottom of the double boiler and place the top over it. Beat the mixture on top of the gently boiling water with the mixer set on high speed for 7 minutes. The icing will have tripled in volume and will hold stiff peaks by this time. Remove the icing from the heat and add the vanilla. Continue beating for a minute to blend the flavor, and then spread immediately on the sides and tops of 2 layer cakes or 24 cupcakes. This icing is particularly good on chocolate cake!

SERVINGS: 2 cakes/ 24 cupcakes

RECIPE FOR: Vanilla Cookies

INGREDIENTS: ½ cup (1 stick) butter or margarine

1 cup sugar

1 egg, beaten

1 teaspoon vanilla extract

1¼ cups all-purpose flour

¼ teaspoon salt

1½ teaspoons double-acting baking powder

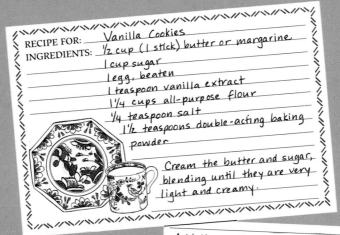

Cream the butter and sugar, blending until they are very light and creamy.

Add the vanilla and the beaten egg. Sift together the flour, salt, and baking powder. Stir the dry ingredients into the butter mixture until it forms a soft dough. Drop teaspoons of dough, leaving generous spaces between them, onto a greased baking sheet and flatten by pressing down lightly with the bottom of a juice glass. (Dust the bottom of the glass with flour or sugar to prevent sticking.) Sprinkle the tops of the cookies with chopped nuts, colored sprinkles, or granulated sugar, if you like, and bake in a 375°F oven for about 10 minutes. Allow the cookies to cool for 5 to 10 minutes before removing them from the pan. Cool

© The Colonial Williamsburg Foundation   C.R.Gibson®, Norwalk, CT 06856

thoroughly before eating. This batter can be refrigerated overnight if you don't have time to bake the cookies right away. Refrigerated dough should be baked at 400° F for 10 minutes.

Makes 2 dozen cookies; recipe can be doubled.

© The Colonial Williamsburg Foundation   C.R.Gibson®, Norwalk, CT 06856

SERVINGS: 2 dozen

Chapter 2

Chocolate:
The Food of the Gods

Is there anything better than chocolate? Rich, smooth, luscious, satisfying chocolate. Almost everyone likes it by itself or prepared in foods: cookies, cakes, pies, hot and cold drinks, ice cream, and candy. Chocolate helps us celebrate. What would Valentine's Day or Easter be without it? It is also the secret ingredient used to improve the flavor of medicines and tobacco, to give color to dark breads like pumpernickel, and to add richness to traditional meat dishes in Italy, Spain, and Mexico.

Many people say that the flavor of chocolate is not only delicious but also comforting. Some people (who jokingly call themselves ''chocoholics'') claim it is addictive. Americans eat almost 11 pounds of chocolate per person per year. That may sound like a lot, but it is only about half the amount that Europeans consume annually.

The native people of Mexico were the first to enjoy chocolate. They learned how to prepare it as a drink and considered the beverage to be ambrosia—the food of the gods.

An engraving from the 1500s shows Aztec women grinding and roasting cacao, techniques the Aztecs learned from Quetzalcoatl.

The Mythical Origin of Chocolate

According to an Aztec legend, a long time ago Quetzalcoatl (pronounced *ket-sahl-ko-AH-tul*), the god of wisdom and knowledge, came down from his land of gold, where the sun rests at night, to be the people's priest king. He taught them how to paint and how to work silver and wood. He gave the people their calendar and showed them how to grow corn. And he brought them the seeds of the cacao tree.

The bearded, white-skinned Quetzalcoatl taught the Indians how to grow the cacao tree, harvest its seedpods, and prepare a delicious drink, chocolatl, from the seeds inside. The Indians continued to prepare chocolatl from the seeds of the cacao tree long after he returned to his land of gold. Before he left, Quetzalcoatl promised he would return to his people. He said he would come in a "one reed" year, which occurs every 52 years in the Aztec calendar.

The European Discovery of Chocolate

When Hernán Cortés arrived in Mexico in 1519, the Aztec king Montezuma II was ruling over a brilliant civilization. The Aztecs thought Cortés, with his beard and white skin, might be Quetzalcoatl returning to his people. As it happened, 1519 was a "one reed" year. Instead of treating Cortés like an enemy, the Aztecs welcomed him with great feasts that concluded with the serving of chocolatl in golden cups.

The chocolatl served to Cortés was not at all like modern chocolate beverages. The roasted, ground seeds of the cacao tree were mixed with cold fermented corn mash and wine or water. The mixture was then flavored with vanilla, hot spices such as pepper and pimiento, and sometimes with a little honey. It was a powerful and heavy drink, not sweet, but bitter and peppery.

An early illustration of the cacao pod

Montezuma and his people treated Cortés like a god, presenting him with a basket of cacao pods (to the left of the column) and other offerings.

This fanciful rendition of Cortés and Montezuma shows the Spanish adventurer bowing to the Aztec leader. In fact, Cortés imprisoned Montezuma and conquered the Aztec nation. Three hundred years of Spanish domination of Mexico and Central America followed. Cortés eventually became the wealthiest person in all of Spanish America, before returning to Spain to spend his final years.

Some historians say that it was intoxicating, or at least that it was sometimes blended with other ingredients that had mind-altering effects.

The Aztecs believed that the drink was an aphrodisiac, or love potion, and that it gave vigor, strength, and wisdom to those who drank it. Montezuma is said to have drunk more than 50 cups daily, and his large household consumed over 2,000 cups a day.

Cortés was not the god that the Aztecs had hoped for. He responded to the people's generous welcome by imprisoning Montezuma, seizing large amounts of gold, and destroying Aztec temples. Eventually, he and his soldiers conquered the Aztec empire and turned their flourishing civilization into an enslaved Spanish colony. When Cortés returned to Spain nine years later, among the treasures he brought with him were cacao beans and the techniques for processing the seeds into a chocolate drink.

Sharing the Treasure

Many of the native groups of Mexico used cacao beans and valued them so much that they served as a type of money. Among the Mayas, 8 beans could be used to buy a rabbit; 100 beans would purchase a slave. When the Aztecs defeated the Maya Indians, they demanded sacks of cacao beans as payment from the conquered people.

Molinets, tools for frothing chocolate drinks, were used first by the Aztecs and later by Spanish and other European chocolate makers.

Cortés fully appreciated the value of cacao beans and considered chocolate to be one of the finest riches that he brought back to the Spanish king, Charles I. The drink became extremely popular in the Spanish court after sugar was added to mellow its bitter flavor. Following the style of the Aztec preparation, the Spanish chocolate drink was at first prepared cold, and thick enough to hold up a spoon. It was beaten into a foam with a utensil known as a molinet, a wooden stick with several loose disks at one end, which was first used by the Aztecs.

Later, when it became fashionable to serve chocolate drinks hot, chocolate pots were made with holes in the lids to accommodate the molinets. The Spanish added spices to their chocolate drinks, including orange water,

Elegant chocolate pots were introduced when it became fashionable to serve chocolate hot rather than cold, as the Mexican Indians had drunk it. Chocolate pots of silver, gold, and porcelain were used throughout the 17th and 18th centuries. Most chocolate pots had holes in the lids to accommodate the molinets.

powdered white roses, cinnamon, musk, nutmeg, cloves, allspice, and aniseed, as well as almonds or hazelnuts. The spices and nuts may have been added to mask undesirable flavors picked up by the cacao beans on the long, damp voyages back from the Americas.

Chocolate and sugar were rare and expensive commodities in the late 1500s. The Spanish royalty wanted to keep these luxuries for themselves. The new drink remained a well-kept secret in the courts of Spain for almost 80 years after it was introduced. Monks in monasteries were entrusted with the task of roasting and grinding the precious cacao beans and shaping the chocolate into little rods or tablets to be used by royal chocolatiers (chocolate makers). Meanwhile, other European countries knew nothing about the new food. In fact, English and Dutch seafarers who captured Spanish trading ships during this time would throw the bags of brown beans overboard.

Monks were the first chocolate makers in Europe. Here, some of them enjoy a cup of their concoction.

In 1606 an Italian merchant named Antonio Carletti brought chocolate from Spain to Italy. From Italy the drink made its way to Austria and Holland. When the Spanish princess Anne of Austria married King Louis XIII of France in 1615, she brought chocolate as a gift for her young husband and introduced the beverage to the French court at Versailles. The shopkeepers of France exported chocolate to England in the mid-1600s. Soon royal chocolate makers across the continent were preparing the drink. By the end of the century, "chocolate houses" had been established in most leading cities, offering the delicious beverage to wealthy citizens along with food, gambling, and lively discussion.

During the 1600s, colonists in the Caribbean and in Latin America became just as fond of chocolate as the native people. Almost everyone in the West Indies and Jamaica drank chocolate. Mexican women liked it so much that they had it served to them in church. When an astonished bishop forbade this practice, he was mysteriously murdered by a poison slipped into his own cup of chocolate!

The North American colonies were slower to adopt the new food. Until the mid-1700s, the chocolate that did make its way to North American settlements was imported from Europe and therefore outrageously expensive. But in 1765, John Hannon, a chocolate maker from Ireland, started a business with Dr. James Baker of Massachusetts to ship cacao beans to America directly from the Caribbean. The popularity of chocolate in America dates from this time.

Throughout the 1600s and 1700s, chocolate grew in popularity and economic importance. Doctors praised and prescribed the drink for the good health and vigor it was said to promote. Wealthy citizens craved it. Governments taxed it heavily. The Aztec drink had captured the taste buds of the world.

Introducing Chocolate to North America

John Hannon, an Irish chocolate maker, came up with the idea of importing cacao beans to the American colonies directly from the Caribbean islands. He received financial backing from Dr. James Baker of Massachusetts, who knew that chocolate was in great demand by apothecaries (druggists). The partners rented space in an old gristmill and ground the roasted cacao beans using water power. While sailing for the West Indies in 1779, Hannon was lost at sea, and Baker took over the company, which still bears his name and produces chocolate products used in baking.

Dr. James Baker

The brightly colored fruits, or seedpods, of the cacao tree grow directly from the main trunk of the tree. The fruits may be red, yellow, gold, pale green, or a mixture of these colors.

The Chocolate Tree

Chocolate is made from cacao beans, which are the seeds of the cacao tree. The scientific name of the tree is *Theobroma cacao*. *Theobroma* means "food of the gods." *Cacao* comes from the Aztec name for the tree's large seedpod, **cacahuatl**. The cacao tree belongs to the Sterculiaceae, a family of plants sometimes referred to as the "cocoa and cola family." This family includes many ornamental plants, as well as the kola trees (*Cola nitida* and *Cola acuminata*) whose nuts originally contributed the characteristic flavor and caffeine found in the world's most popular soft drink.

Scientific Names, Scientific Relationships, and Mistaken Identities

The botanical or scientific name of a plant consists of two Latin words. The first word indicates the genus of the plant and the second word indicates the plant's species. For the cacao tree, *Theobroma* is the genus name and *cacao* is the species name. In modern botanical classification systems, plants that have flowers and fruits with similar characteristics are grouped together into the same genus. For instance, all trees that bear fruits known as acorns and have separate male and female flowers are grouped together into the genus *Quercus*. We call them oak trees. But there are many types of oak trees: black oaks (*Quercus kelloggii*), valley oaks (*Quercus lobata*), blue oaks (*Quercus douglasii*), scrub oaks (*Quercus dumosa*), and so forth. They differ from one another in shape of leaf, areas where they grow, and growth habits, but they can interbreed with one another to produce hybrid offspring. They are different species of the genus *Quercus*; that is, subdivisions of the group of plants known as oaks.

Just as closely related species of plants make up a genus, related genera (plural of genus) of plants are grouped by botanists into plant families. For example, oaks, beeches, and chestnuts are grouped together into the beech tree family or *Fagaceae*.

Some people think that the cacao tree is related to cocaine or coconut. Although the common names and the botanical names may sound somewhat alike, *Theobroma cacao* is not at all related to the coca plant, *Erythroxylon coca*, the plant from which cocaine is derived. Nor is it related to *Cocos nucifera*, the coconut palm. These three plants are not only in different genera, they are in different families of plants as well.

Some of the well-known plants in the "Cocoa and Cola" family include the Fremontodendron, a bush native to California named after John Frémont, the explorer and early senator of California, and Brachychiton, the ornamental Australian flame tree. Kola trees, originally used in making well-known soft drinks, are also members of this family.

cacao tree

coca plant

coconut palm

Illustrations not drawn to scale.

Cacao trees grow wild in the tropics of the Americas. Most of the trees have been found within 10° north or south of the equator.

The cacao tree can be cultivated in many different regions, as long as the land is no higher than 3,000 feet in elevation and the temperature is approximately 80°F year-round and never falls below 60°F. The tree thrives on abundant and steady rainfall—at least 50 inches per year—and high humidity. It grows best in loose soil in areas where there is little wind. Wild cacao trees usually grow in the shade of larger trees in the South and Central American forests, and cultivated trees do best in screened sunlight.

In their natural state, cacao trees grow as tall as 50 feet and live about 60 years. Trees grown on plantations, however, are kept pruned to about 25 feet to make the gathering of the bean pods easier. Plantation trees are kept in production for 30 to 40 years before they are replaced with younger, more productive trees.

The cacao tree is a beautiful evergreen plant with large, dark green leaves and fragile branches. The flowers and fruit do not grow from the branches as they do on most trees, but instead grow directly from the trunk or main branches. Cacao flowers are small, about a half inch in diameter, and pink or white. They grow in small clusters on the trunk of the tree.

Each five-petaled flower has both male and female reproductive parts, but no nectar or scent to attract pollinators. The pollen is too sticky to be distributed by the wind. Scientists think that the main pollinator of the tree is a little mosquito-like fly called a midge. This pollinator is not very efficient, though, because less than 5 percent of the flowers produce fully developed pods.

After cacao flowers are pollinated, the fruits that develop from the flowers are shaped like footballs, about 10 inches long and 3 to 4 inches around. Each fruit,

The pulp and seeds of a cacao pod. Inside the pod, the seeds are stacked in five columns around a central core. Each seedpod usually contains 20 to 40 almond-shaped seeds, or beans.

called a pod, contains 20 to 40 seeds that look like shelled almonds. The pods take about five to six months to mature. As they ripen, their color changes from green to yellow or purple. Inside the fruit, the purple seeds are embedded in a sweet, white, pulpy substance.

Since the cacao tree flowers practically every day of the year, there are always fruits on the trunk, their various colors reflecting their different states of maturation. A cacao plantation is a beautiful sight all year round. But it is a feast for the eyes only, because neither the flowers nor the fruit give off even a hint of the mouthwatering smell of chocolate.

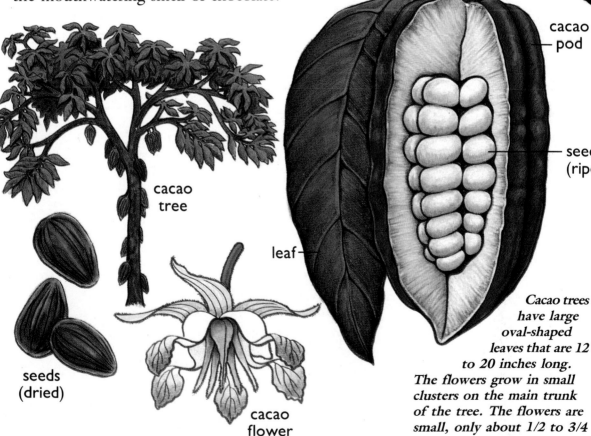

cacao pod

seeds (ripe)

cacao tree

leaf

seeds (dried)

cacao flower

Cacao trees have large oval-shaped leaves that are 12 to 20 inches long. The flowers grow in small clusters on the main trunk of the tree. The flowers are small, only about 1/2 to 3/4 inch across, and have 5 petals.

Growing Cacao on Plantations

On cacao plantations, young trees are raised from seed in a nursery until they are four to six months old. Then they are planted 7 to 10 feet apart near "mother plants" such as banana, cassava, or maize, which will provide the cacao trees with shade and protection from wind. The trees begin bearing fruit when they are three years old. Their yield of fruit increases over the next 5 years and then continues at a steady level for about 30 more years. Each tree produces between 20 and 50 pods per year, enough to make 5 to 13 pounds of processed chocolate.

On tropical plantations, cacao beans are usually harvested twice a year, before and after the rainy season. The trees are never picked bare of all their pods. Only the ripe fruit is harvested.

Cacao trees were being cultivated by Indian farmers in the Yucatán Peninsula in Mexico long before the arrival of Cortés. When the Spanish took over the rule of Central America, they also took over its agriculture and trade and went on to establish many more cacao plantations, first in Mexico and Venezuela, and later in Fernando Póo (an island off the coast of West Africa) and the Philippines.

For almost a century, the Spanish had a monopoly on cacao production. But as the demand for chocolate increased across Europe, other nations began to establish cacao plantations. The Dutch started operations in Ceylon (now Sri Lanka), Java, and Sumatra; the Belgians in the Congo and central Africa; the English in the West Indies; the Germans in Cameroon; and the French in Martinique and Madagascar. Currently, the countries of West Africa, including Ghana, Côte d'Ivoire, Nigeria, and Cameroon, lead the world in cacao production. South American countries such as Brazil, Colombia,

A cacao tree

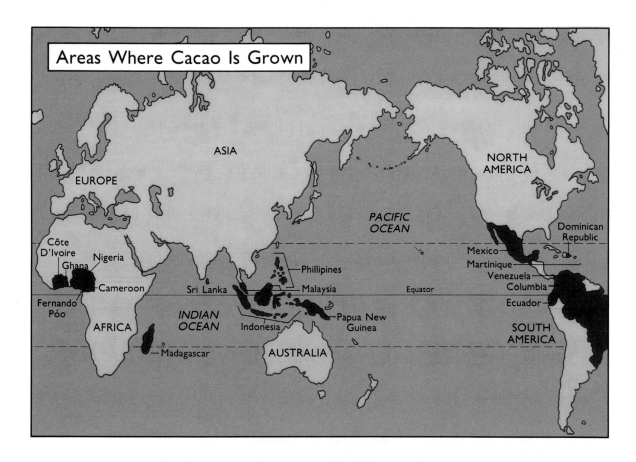

Areas Where Cacao Is Grown

Venezuela, the Dominican Republic, and Ecuador harvest the second largest crop. Malaysia, Indonesia, and Papua New Guinea are also significant producers. Mexico, the plant's land of origin, contributes only about 2 percent of the total world crop.

Cacao is grown on plantations in Africa, Central and South America, and the Far East. The countries involved in cacao production are indicated on the map.

Curing Cacao Beans

Cacao beans, the seeds of the cacao tree, must be cured (fermented and dried) to bring about the chemical changes that produce the delicious smell and taste of chocolate. The process for curing chocolate is similar to that used for vanilla. First, plantation workers cut the

The first step in the making of chocolate is harvesting the cacao pods.

ripe pods from the tree trunk, being careful to avoid injuring the young pods still ripening on the tree. Next, the tough pods, which weigh about a pound each, are split open with a stick or knife. At this point, the seeds themselves are very bitter, but the sticky white pulp surrounding the beans is sweet and delicious. People who live in cacao-raising areas often eat the fresh pulp.

A Nigerian woman removes cacao beans from pods.

It is difficult to remove the clinging pulp from the fresh beans, so both are scooped out of the pods together. Traditionally, the pulp and beans were piled on the ground and covered with leaves to ferment, but in modern plantations, the beans and pulp are heaped into wooden boxes (which have drainage holes in the bottom), covered with lids, and laid in the hot sun. Temperatures in the boxes reach as high as 110°F. This heat causes the beans to start to grow and the pulp to ferment. Enzymes trigger the conversion of sugars in the pulp into alcohol and carbon dioxide. Then bacteria in the pod change the alcohol into acetic acid. The sticky pulp becomes liquid and drains away from the beans in about a day.

Meanwhile, the beans themselves are undergoing important changes. As the heat in the mound of fermenting cacao beans continues to rise, reaching about 122°F,

After the cacao beans have fermented, which takes about a week, they are laid in the sun to dry, just as vanilla pods are.

the seeds, which had started to grow, die. They change color from creamy white to lavender to a dark brown—the color we normally associate with chocolate. The outside coat of each bean hardens. The taste of the beans mellows, and compounds form that will eventually develop into the chocolate taste. These compounds are called **flavor precursors.**

The whole fermentation process lasts 5 to 7 days. After the cacao beans have fermented, they still contain quite a bit of water, sometimes as much as 60 percent of their weight. So the beans are spread out to dry under the hot tropical sun until their moisture content is between 5 and 8 percent. As they dry, enzymatic action continues within the beans. The seeds turn darker brown and more flavor precursors develop. When the beans are dry enough, they are shipped to chocolate factories around the world for processing.

The dried cacao beans (top) *are packed in sacks and shipped throughout the world for processing.*

Processing Chocolate

If cacao beans are not cured properly, they will never develop the rich flavor and aroma we associate with chocolate. But curing is only the first step in the preparation of chocolate products. The processing of the cured beans is also important in the formation of the flavor. This processing usually takes place in giant factories.

When cured cacao beans arrive at a factory, they are first cleaned. Debris such as stones, sand, or wood particles are removed by passing the seeds through electric brushes, vacuums, and sieves. Then the beans are separated according to size. The bigger the beans, the longer they must be roasted.

Roasting is the most important step in the processing of cacao beans into chocolate. The roasting time and temperature must be controlled very carefully, for it is during this procedure that the enticing chocolate flavor

The cacao beans are roasted in large cylinders or on a conveyor belt. In a cylinder, the beans are turned as they roast to prevent burning.

and smell fully develop. If the beans are underroasted, they will have an inferior taste. The longer the beans are roasted, the darker their color becomes and the stronger their flavor—but overroasting can completely ruin a whole batch of beans. On average, cacao beans are roasted at a temperature of between 225 and 300°F for 15 to 20 minutes. As they roast, the beans lose even more of their water content and become dry and brittle.

After roasting, the cacao beans are cracked open by a winnowing machine. The seed coats of the cacao seeds are blown off; they will be used in cattle feed or fertilizers. The seeds themselves, now referred to as **nibs,** are ready to be ground. Grinding is done by machines that crush the beans between heavy disks of stone or stainless steel. The grinding, or milling, releases the fat from the cells of the cacao beans. The grinding generates enough heat that the fat, called **cocoa butter,** melts.

The crushed nibs and melted cocoa butter form a chocolate-smelling paste, which hardens as it cools. This rich, dark liquid is known as **chocolate liquor** ("liquor" means that the material is liquid, not that it contains alcohol). Chocolate liquor is the basic material from which other chocolate products are made.

When chocolate liquor hardens, it is pure chocolate—as pure as you can get it. Cooled, solidified chocolate liquor is sold in stores as baking chocolate. If you taste a piece of pure baking chocolate, you will notice that it is bitter and rich, since it is made up of more than 50 percent cocoa butter. Hot chocolate made from baking chocolate is very different from cocoa, the beverage that we commonly drink. The process for making cocoa, which has much less fat, was invented by a Dutch chemist by the name of Coenraad van Houten in about 1810.

Van Houten discovered a way to remove about two-thirds of the fat in chocolate liquor. Using a press he invented, he extracted cocoa butter from molten chocolate liquor. He found cocoa butter to be a yellowish fat with a pleasant taste and aroma, which kept well and was easily handled. Unlike most vegetable fats, such as corn oil, cocoa butter is solid at room temperature but melts at 92 or 93°F, the normal temperature of the human tongue.

A hard cake of chocolate material was left behind after the cocoa butter was removed from the chocolate liquor. Van Houten pulverized this cake into an extremely fine powder and called it cocoa. The less fatty product was much easier for people to digest. Moreover, it dissolved in water more readily and was easier to work with in making candy and cakes. The chemist packaged this pulverized material as breakfast cocoa and sold it to bakers, candy makers, and housewives.

Cocoa for Chocolate

Cookbooks often suggest that three tablespoons of cocoa plus one tablespoon of vegetable oil can be substituted for one ounce of baking chocolate in recipes. This is a way of making chocolate liquor by adding back fat that was removed in the manufacture of cocoa. Vegetable oil is liquid at room temperature, however, and makes a poor substitute for rich cocoa butter. The substitution trick works in recipes, but the chocolate liquor you make in this way will never give real chocolate any competition.

*Coenraad van Houten,
inventor of cocoa*

Cocoa remains the most common chocolate flavor ingredient in everything from syrups, cookies, and puddings to pharmaceutical products. Unsweetened cocoa is sold in most grocery stores. You might have it in the cupboard of your kitchen, because it is an ingredient often called for in making chocolate cakes and frostings. Taste a little of it so you can see what pure chocolate flavor, without sugar and vanilla, is really like.

Sometimes at the store you will see a type of cocoa labeled "Dutch processed." This was another process discovered by van Houten, called "dutching" because its inventor was from Holland. In the 1800s, when chocolate was not quite as dark as the customers wanted it, store owners took to adding some brick dust or other red coloring to deepen the color (a questionable practice!). Van Houten discovered that by treating the nibs or the chocolate liquor with an alkaline material such as sodium bicarbonate (baking soda), he could deepen the color without harming the cocoa. The process does affect the taste of the cocoa somewhat, though. Taste

*Hard cakes of cocoa are
broken up, then smashed
into a fine powder.*

some "Dutch processed" cocoa if you get the chance and see if you can detect the difference.

Making Molded Chocolate

When van Houten started taking the cocoa butter out of chocolate liquor, he was left with a lot of pure cocoa butter. It turned out to be a very good base material for cosmetics like lotions and soaps. More importantly, it was discovered that if some of this extra cocoa butter was added to chocolate liquor along with sugar, chocolate could be molded into a form that was solid at room temperature, but melted on the tongue. In other words, the chocolate bar was born! The first chocolate bars were made in the 1840s by the Englishmen John and Benjamin Cadbury and Joseph Fry.

To make molded chocolate—which is used in candy bars—the right amount of cocoa butter and sugar must be added to chocolate liquor and mixed thoroughly. The resulting mixture is often a bit gritty. To achieve a smoother texture, the mixed batch is sent to a **refining** machine, where steel rollers rotate over and crush the tiny particles of chocolate and sugar. But even this refining isn't quite enough. For the best texture, the chocolate must go through a process known as **conching.**

Rodolphe Lindt invented the conching process and founded the company that still bears his name.

Conching was invented by Rodolphe Lindt in Switzerland in 1893. He discovered that kneading chocolate in a trough gave it a smoother, more pleasing texture. He invented machines to knead and mix the chocolate after it is refined. The process is called "conching" because the troughs originally used to knead the chocolate were shaped like conch shells. Sometimes extra cocoa butter is added to chocolate during conching, but it is the process itself that harmonizes all the flavors.

From the production of chocolate liquor through the conching process, the chocolate is kept in a liquid state.

After the final conching is completed, the chocolate must be cooled carefully to prevent the cocoa butter from recrystallizing, which would make the finished chocolate streaky and grainy. In a process known as **tempering,** chocolate is gradually cooled from temperatures as high as 185°F. The tempered chocolate is then placed into pouring machines, which fill solid molds for chocolate bars, line molds that will be filled with creams and topped with more chocolate, and shower chocolate onto chewy centers. Then the chocolate bars and candy are ready to be packaged and shipped to stores.

Facing page: In the conching machine, the chocolate is heated and a roller travels back and forth, stirring the chocolate. Below, bottom: Before the conching comes the refining. After passing through the refining, conching, and tempering stages, the chocolate is ready to be molded into the familiar shapes of candy, such as kisses or bars.

Great Names in Chocolate

Many famous brand names of chocolate commemorate people who made major contributions to chocolate manufacturing. Most chocolate companies were started in the last 100 years.

James Baker A Massachusetts physician who in the mid-1700s started the first chocolate manufacturing company in America, along with the Irishman John Hannon.

Coenraad van Houten A Dutch chemist who invented the process and the press to remove cocoa butter from chocolate and produce cocoa powder. Also developed the "Dutch processing" of cocoa to improve its color.

John and Benjamin Cadbury Started manufacturing chocolate in England in the mid-1800s, and along with Joseph Fry made the first chocolate bars in the 1840s. When the brothers, who were staunch Quakers, learned that the beans they purchased were grown on plantations that used slave labor, they discontinued business with them and supported cacao plantations that employed native people on mainland Africa.

Henri Nestlé A French Swiss who learned how to condense milk, an important step in the manufacture of milk chocolate, which was first successfully produced by Daniel Peters.

Rodolphe Lindt

A German Swiss who invented the conching process to improve the texture of refined chocolate.

Milton Hershey

An American who started the famous candy company in Pennsylvania in 1894. A Mennonite, he opened a school for orphan boys in 1909 that was and continues to be supported by the profits of the company.

Jean Tobler

A Swiss citizen who started manufacturing chocolate in that country in 1899.

Milton Hershey not only started a chocolate factory but also built a town. Hershey, Pennsylvania, included restaurants, housing, businesses, and schools. The town soon became a tourist attraction and continues to draw many visitors every year.

Kinds of Chocolate

You already know the difference between cocoa and baking chocolate (hardened chocolate liquor), but it can be confusing to figure out the differences between other types of chocolates.

Milk chocolate was invented in 1876 by Daniel Peters of Switzerland. Milk chocolate, as you might guess, is chocolate with sugar and milk added to it. Henri Nestlé's earlier invention of condensed milk contributed to the creation of milk chocolate. Mixing regular milk into chocolate results in a mixture that is too runny to mold well. Nestlé's condensed milk, however, could be dried out to the consistency of cream cheese. It worked well for creating a moldable chocolate with a distinct taste.

The Swiss government requires that milk chocolate be at least 25 percent chocolate liquor, 14 percent milk, and no more than 55 percent sugar. In the United States, though, milk chocolate can have as little as 10 to 15 percent chocolate liquor. (Governments establish standards of identity for many products so that consumers can expect consistency in the items they buy.)

Semisweet chocolate is chocolate made without milk, but with a moderate amount of sugar. In the United States, it usually contains about 35 percent chocolate liquor. Sweet chocolate is also chocolate made without milk, but with more sugar. Sweet chocolate contains about 15 percent chocolate liquor. Both of these chocolates are used to make chocolate-covered candy, cookies, and cakes. You can look at the labels on packages of chocolate chips at the grocery store to see what kind of chocolate they contain.

The amount of extra cocoa butter that is added to chocolate increases as you go from bittersweet to sweet

The inventor of condensed milk is a famous name in the world of chocolate.

and from dark to light varieties. The greatest amount of cocoa butter is found in white chocolate, which, strictly speaking, contains no chocolate at all—it is a mixture of cocoa butter, milk, and sugar.

Imitating Chocolate

A huge number of food products sold today are chocolate flavored, but only a few contain pure chocolate. It can be quite difficult to read the list of ingredients on the side of a package and understand exactly what is in the food. For instance, "compound chocolate" does not mean that the chocolate has been made from different types of beans, or mixed from different processing batches. It means that the manufacturer has completely or partially substituted other vegetable fats for the cocoa butter in the chocolate. Real or genuine chocolate must be made with 100 percent cocoa butter, but compound chocolate can contain a variety of different fats and oils. Both genuine and compound chocolates often have emulsifiers added to them to keep the product well mixed but still soft. The most common emulsifier added to chocolate is lecithin, a nutritious substance derived from soybeans.

Chocolate products such as cake mixes, ice cream, puddings, syrups, and toppings often contain chocolate or cocoa. Cocoa, either plain or Dutch processed, is used most often as the chocolate flavoring ingredient in processed foods. You can confirm this fact by spending a few minutes looking at the ingredient lists on a variety of products on your next trip to the grocery store. (Hot cocoa mixes and chocolate-flavored drinks, by the way, are not the same as pure cocoa. Although these drink mixes contain a small amount of cocoa, they also contain milk powders, preservatives, flavorings, and a lot of sugar.)

Legally, products cannot be called "white chocolate" in the United States. The U.S. Food and Drug Administration sets standards of identity specifying the ingredients products must

contain in order to be labeled "chocolate." Chocolate products must contain chocolate liquor. What we call white chocolate is made of cocoa butter, sugar, and milk— and no chocolate liquor.

Many products with "chocolate flavor" contain little if any real chocolate.

Sometimes chocolate extract or artificial chocolate flavoring is used in packaged foods. Chocolate extract is made by mixing cacao nibs or cocoa with alcohol and then **distilling** the mixture to concentrate the flavor. Artificial chocolate flavoring, on the other hand, is a synthetic potpourri of chemicals. These artificial mixtures usually only hint at the flavor of chocolate and are often darkened with coloring agents. Both chocolate extract and artificial chocolate flavoring are poor substitutes for the real thing. They are used by bakers and candy makers, usually to supplement the cocoa or chocolate in a recipe rather than to completely replace either one.

When a product is labeled "chocolate-flavored" instead of "chocolate," it usually contains some chocolate or cocoa, but not enough to meet the minimum standards set by the federal government. Such chocolate-flavored products typically contain artificial chocolate flavoring or chocolate extracts.

The flavor of chocolate is hard to duplicate chemically. Scientists have identified over 200 compounds that contribute to the complex flavor of chocolate. These compounds are often referred to as "volatile," because they escape into the air easily and reach the nose (as a good smell). But even if all these volatile compounds are removed, the remaining material still has the main characteristics of chocolate, although milder. This is because much of the flavor and aroma of chocolate is locked in the food's basic nutrients—its sugars and **amino acids**. When the cured cacao beans are roasted, complex mixtures of sugars and amino acids, called flavor molecules, are formed. These and the volatile compounds all contribute to the unique taste of chocolate!

Some people feel that other foods taste similar to chocolate and use them to manufacture products with an imitation chocolate flavor. One imitation chocolate flavor is a powder made from carob, the fruit of the locust tree *Ceratonia siliqua,* which is sometimes called St. John's bread. This powder duplicates the color and texture of cocoa, but certainly not the flavor. Another natural imitation chocolate flavoring is a mixture of processed barley and roasted malt extract. Although imitation chocolate flavors are not unpleasant, few people have difficulty distinguishing them from the taste of genuine chocolate.

The Nutritional and Medicinal Value of Chocolate

Chocolate is more than just a flavor. It is a food that packs a nutritional punch. A 3.5-ounce chocolate bar has about 500 calories, as much as a third of a pound of cheese or 2 pounds of bananas, and a little less than

Caffeine and Theobromine

A look at the chemical structures of caffeine and theobromine shows you that the two compounds are very closely related. In caffeine, a methyl group (CH_3) replaces the hydrogen (H) attached to the nitrogen (N) in the number 6 position on the benzene ring. This small chemical change makes a big difference in the effect the compounds have in the human body. They are both stimulants, but caffeine affects the nervous system, while theobromine affects only muscles.

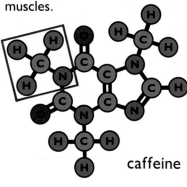

caffeine

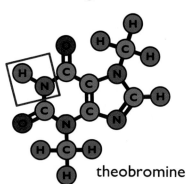

theobromine

half a loaf of bread. Chocolate is very high in carbohydrates, which makes it a fast energy food. For that reason, sugar-sweetened slabs of chocolate were issued as a ration to British seamen from 1780 until 1968. In the mid-1800s, approximately half of the cacao imported into Great Britain was allotted to the Royal Navy. Similarly, during World War II, the Hershey Company developed a 600-calorie chocolate bar, known as "the Field Ration D bar," which could be used as a survival diet.

Chocolate is also rich in fat, which is digested more slowly than proteins or carbohydrates, providing energy over a longer period of time. Climbers and adventurers often carry chocolate in their packs for both its nutritional and psychological value. In the early 20th century, chocolate was issued as a ration for the antarctic expeditions. The members of Roald Amundsen's successful trek to the South Pole drank nothing but chocolate for 99 days!

Chocolate is a good source of B vitamins and an excellent source of minerals, especially calcium, phosphorus, iron, and copper. Like tea and coffee, chocolate contains caffeine. It also contains a related compound called **theobromine.** But a cup of cocoa has only half the stimulants found in a cup of coffee, and a chocolate bar has amounts too small to be noticeable.

Although early Europeans believed in the health benefits of chocolate, its most significant medical use today is to make medicines taste better. Theobromine is a muscle stimulant, but in the amounts normally eaten in cocoa and candy, it has little or no effect. Despite its long association with love, there is no strong chemical evidence to validate claims that chocolate is an aphrodisiac. Phenylethylamine, however, one of the compounds in chocolate, is known to play a role in the perception of pleasure.

Chocolate can have an irritating effect, unfortunately, on people sensitive to it. Some people find it gives them severe headaches. For some unlucky individuals, chocolate triggers outbreaks of the allergic rash known as hives.

For most people, though, chocolate is the most satisfying of all indulgences, a perfectly "divine" food. Following are some recipes for chocolate in its historic and delicious forms.

RECIPE FOR: Aztec Chocolate and Spanish Chocolate
INGREDIENTS: 1 ounce unsweetened baking chocolate
1 teaspoon vanilla
2/3 cup boiling water
ground pepper or chilies to taste

Aztec Chocolate
Grate the unsweetened chocolate
into a bowl and cover it with a
little of the boiling water. Mash
the mixture into a paste.
Add the rest of the water
and vanilla and beat with
an electric mixer until frothy.

The mixture can also be beaten in a blender set on high
speed. Add pepper or chilies to liven up the
drink a bit!
The chocolate does not totally dissolve in the water
using this technique. Tiny particles of chocolate will
float in the water, and you will be able to taste the
grittiness in the drink. For a more authentic drink,
allow the chocolate to cool and beat it into a froth just
before you drink it.

Spanish Chocolate
Follow the directions for the preparation of Aztec chocolate, but omit the chilies and pepper and add 3 teaspoons

of sugar, plus a dash of cinnamon or nutmeg if desired.
The sugar makes a critical difference in the taste of
this rich but delicious beverage.

SERVINGS: 1

RECIPE FOR: Microwave Chocolate Fudge

INGREDIENTS: 8 ounces semisweet chocolate
2/3 cup sweetened condensed milk
1 teaspoon vanilla extract
1/8 teaspoon salt
1/2 cup chopped nuts (optional)

Break the chocolate into a 1½-quart
glass or microwave-safe bowl. Add
the milk and microwave for one
minute. Stir well and microwave the mixture on high for
another minute. Stir until
the chocolate is completely

melted and smoothly mixed into the milk. Stir in the
vanilla extract, salt, and nuts. Spread into a greased
loaf pan and refrigerate until firm. Cut into small
squares before serving.

SERVINGS: 8-10

© The Colonial Williamsburg Foundation   C.R.Gibson®, Norwalk, CT 06856

RECIPE FOR: Double Chocolate Brownies

INGREDIENTS: 1 cup granulated sugar
1/2 cup vegetable oil
1 teaspoon vanilla
2 eggs, beaten
1/2 cup flour
1/3 cup pure unsweetened cocoa
1/4 teaspoon baking powder
1/4 teaspoon salt
1/2 cup chopped walnuts
1/3 cup semisweet chocolate bits
Powdered sugar (optional)

Preheat the oven to 350°F. Blend

the sugar, oil, and vanilla in a mixing bowl. In another
small bowl, beat the eggs and add them to the
sugar mixture, beating well to mix. Combine the
flour, cocoa, baking powder, and salt. Gradually
stir the dry ingredients into the egg mixture. Add the
nuts and chocolate bits. Spread mixture in a greased
9-inch baking pan. Bake at 350°F for approximately
30 minutes, or until the dough starts to pull away from
the edges of the pan. Cool thoroughly in the pan.
Sift powdered sugar over the brownies, if desired,
and cut into about 16 squares. These are easy to make
and delectable to eat!

SERVINGS: 16

© The Colonial Williamsburg Foundation   C.R.Gibson®, Norwalk, CT 06856

Chapter 3

Strawberry:
The Scattered Fruit of a Rose

When the weather warms and the days lengthen, the ripening of strawberries in woods and meadows, gardens and farms heralds the arrival of spring. Around the world, people celebrate the start of the growing season by serving the first luscious berries of the season. Scandinavians blend them with yogurt and cream into a sweet, cold soup. Other Europeans use them in fresh fruit tarts. Americans slice them onto their breakfast cereals, cook them into preserves, and pile them high on shortcakes, with lots of whipped cream.

Strawberries grow wild in regions around the world where the climate is mild, including Europe, North America, and the western coast of South America. It's likely that the fruit has been collected and eaten by humans since prehistoric times, but the first evidence that the wild berries were enjoyed at meals comes from accounts of Roman banquets in about A.D. 50.

The Romans referred to strawberries as *fraga*, meaning "fragrant." The first part of the botanical name of the plant, *Fragaria*, is derived from *fraga*, as are the common names for the fruit in Italian (*fragola*), Spanish (*fresa*), and French (*fraise*).

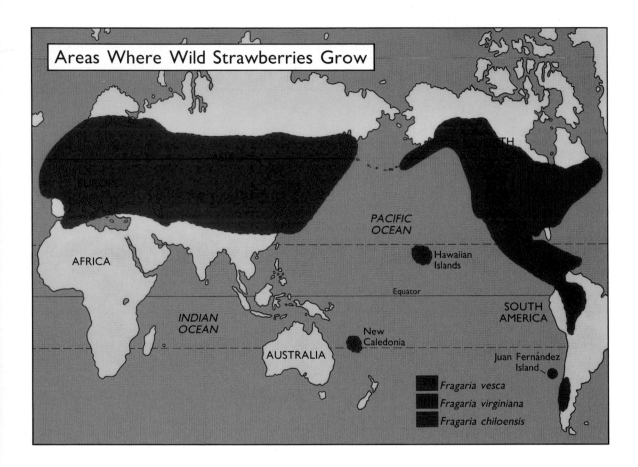

Areas Where Wild Strawberries Grow

ASIA

EUROPE

NORTH AMERICA

PACIFIC
OCEAN

AFRICA

Hawaiian
Islands

Equator

INDIAN
OCEAN

SOUTH
AMERICA

New
Caledonia

AUSTRALIA

Juan Fernández
Island

Fragaria vesca
Fragaria virginiana
Fragaria chiloensis

Wild strawberries grow across the world in moderate climate zones. The areas where the wood strawberry (Fragaria vesca), *the Virginia strawberry* (Fragaria virginiana), *and the Chilean strawberry* (Fragaria chiloensis) *grow are shown on the map.*

It is unclear how the fruit became known as "strawberry" in the English language. In Anglo-Saxon (the language from which English evolved), the word *streow* meant "hay." Some scholars think that the fruit was called "streowberry" (hayberry) because it ripened at the time the hay was mown. Others suggest that the name came from the fact that the fruit was "strewn" or scattered among the leaves of the plant, or that the long **runners** produced by the plant were "strewed" over the ground (a word often spelled as "strawed" in Anglo-Saxon). The name could have arisen from the common practice (still followed) of placing straw under the ripening fruit to protect it from insects and snails in the soil. It is even possible that the name originated from the custom among London children of threading the berries on grass straws and selling "straws of berries."

No matter what the origin of the fruit's common English name, one thing is certain: the wild berries collected by early Europeans were quite different from the big, juicy fruits we enjoy today. The development of the modern strawberry did not take place overnight, nor was it accomplished by one person. It resulted from a series of steps involving the **domestication** of wild plant species, political intrigue, the unraveling of a botanical mystery, and extensive breeding programs. The quest to produce better strawberries has been a continuing scientific adventure.

Domesticating the Strawberry

To domesticate a plant means to collect it (or its seeds) from an area where it grows naturally and to find the right conditions for it to grow and reproduce in a garden setting for human use. No one knows exactly when people began to dig up strawberry plants from the wild and raise them in their own gardens, but by A.D. 1300 it was a common practice. In the middle of the 1300s, King Charles V of France ordered 12,000 strawberry plants to be collected and set out in the French royal gardens. By the 1500s, garden-grown strawberries were popular enough to be sold in London markets. Shakespeare even mentioned them in one of his plays. Books on gardening provided instructions on how and when to transplant strawberries from the wild. Botanical books, known as herbals, included drawings of the plant along with accounts of its medicinal uses.

European gardeners transplanted the healthiest wild plants, bearing the biggest berries, into their gardens.

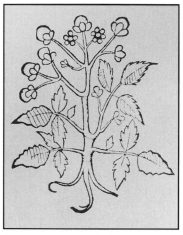

During the Middle Ages, strawberries were believed to have medicinal properties. They were usually included in herbals, books that described plants useful for healing. The first printed illustration of a strawberry (above) appeared in 1484 in an herbal called the Herbarius Latinus Morguntiae. *The book was printed by Peter Schöffer, the partner of Johannes Gutenberg.*

Left: Fragaria vesca, *the common wood strawberry*. Right: Fragaria virginiana, *the Virginia strawberry*.

Still, wild European strawberries bore small fruit, and the strawberries eaten in Europe before 1600 were about the size of a small grape.

Discovering New Strawberries

By the mid-1500s, it was well known that several different types of strawberry plants grew wild in Europe. The delicate plant normally transplanted into gardens for cultivation was the common wood strawberry (*Fragaria vesca*). It had small leaves and flowers and bore its fruit close to the ground. Another type of strawberry, known today as the musky-flavored strawberry (*Fragaria moschata*), was also cultivated. The fruit of this plant was bigger, but also softer and pastier in texture, and it had a musky flavor. Its flowers and fruit grew on long stems that rose high above the plant's rough leaves.

In the 1600s, when European colonists first settled on the East Coast of North America, they found vast patches of strawberries growing in the meadows and along riverbanks. The fruits borne by these strawberry plants, now known as the Virginia strawberry (*Fragaria virginiana*), were wonderfully fragrant and had a strong, sweet, but acidic flavor. They were also three to four times larger than their European relatives. The colonists domesticated this strawberry and cultivated it in their gardens.

When European botanists read about the big-fruited Virginia strawberry, they were anxious to grow the plant themselves. They brought the Virginia strawberry back to England and France in the 1600s, but the plant did not grow well in the climate and soil there.

Meanwhile, Spanish settlers arriving on the western coast of South America discovered that the Indians there were already cultivating a type of strawberry that grew wild in the area. Now known as the Chilean strawberry (*Fragaria chiloensis*), the leaves of the plant were thicker and stiffer than those of its relatives. One petal of the Chilean strawberry flower was as big as the whole blossom of the European wood strawberry, and the fruit was as big as a walnut, sometimes as big as a hen's egg. Since the Indians had two different words for the wild strawberry and the domesticated plant, the Spanish settlers concluded that the native people of South America must have been cultivating the plant for a long time.

Fragaria chiloensis, *the Chilean strawberry*

The Chilean strawberry was not introduced to Europe until 90 years after the Virginia strawberry had arrived. What's more, it was not brought back by horticulturists, but by a French spy! In 1712 a young engineer named Amédée François Frézier was sent on a secret mission by King Louis XIV of France. Posing as the captain of a merchant vessel, Frézier was to examine the defenses installed by the Spanish at their colony in Concepción,

The Curious Origin of Frézier's Name

Like many educated men of his time, Amédée François Frézier was interested in botany—the study of plants. He may have had a particular interest in strawberries because of his surname. Frézier is an ancient name derived from the French word for strawberry (*fraisier* means strawberry plant in French). Legend has it that Frézier had an ancestor by the name of Julius de Berry, who lived in Auvers. In late May of A.D. 916, the emperor king of France, Charles Simplex, stopped in Auvers en route from Lyons, where he and Cardinal Clemens of Monte Alto, Italy, had gone to settle some disputes. The emperor ordered a sumptuous feast prepared for his guest, the cardinal. At the end of the entertainment, Julius de Berry presented the men with dishes of ripe strawberries. The emperor was so delighted with the offering that he knighted Julius, changed his name from Berry to Fraise (which later evolved into Frézier in France and Fraser or Frazer in the British Isles), and gave the family three strawberry blossoms as their coat of arms.

Chile. While Frézier's ship was anchored in the Chilean port, the Spanish authorities thought he was collecting goods to take back and sell in France. Instead, he was busy making detailed observations that would help the French army attack the city—information about fortifications, approaches for attack, the size of the army, the number of guns, and the location of ammunition.

Among the things that Frézier's keen eyes noted during this stay was a thick-leaved strawberry plant with enormous flowers and fruits. The fruit tasted a little like pineapple and was not fully red when ripe, but yellowish and pale at the base. The fruits were not completely round, but long and angular. Instead of hanging down like other strawberries, they pointed upward on the plant so that the end of the berry faced the sun. Although descriptions of such strawberries had been reported by other Europeans traveling in South America, no one had ever brought the plant back to the continent.

Frézier saw this plant growing wild. He also saw it being cultivated in fields outside the city. He was so taken with the plant that he carried five living specimens with him on the six-month journey home to France. He made sure they received enough moisture to survive, despite the very limited rationing of fresh water on the ship.

Unlike the Virginia strawberry, the Chilean strawberry grew vigorously in Europe, particularly in the north coastal regions of France. But Frézier's new plant was hardly an instant success. Although it grew well, it produced no fruit. For 50 years, no one could figure out why. It seemed that North and South American strawberries, which held so much promise, were destined to fail when grown outside their home areas.

Horticulturists were disappointed. Still, a very important step had been taken in the development of the modern strawberry. The two New World strawberries, which grew naturally thousands of miles apart, in Chile and Virginia, had finally been brought together on a continent foreign to both of them. In Europe, where scientists could study them more closely, the two plants would eventually become the parents of the modern strawberry.

Frézier sketched the strawberry he found in Chile, which, he noted, was "as big as a walnut and sometimes as a hen's egg."

Unraveling a Mystery

On July 6, 1764, Antoine Nicholas Duchesne, the young son of the superintendent of buildings at the palace of Versailles, personally presented the French king, Louis XV, with a potted plant of strawberries in full fruit. It was a Chilean strawberry plant, and its fruit was extraordinarily large and beautiful. The 17-year-old boy had succeeded in unraveling a mystery that had haunted strawberry growers for half a century.

Duchesne had been doing some experiments on the European musky-flavored strawberry, which was considered very difficult to raise. While almost every plant in a plot of wood strawberries would bear fruit, gardeners frequently found that half the plants in a plot of the musky-flavored variety produced no fruit and had to be torn out. But when this was done, the remaining plants would bear less and less fruit. Soon the whole plot would have to be replaced by new, more vigorous plants.

Duchesne knew that the European wood strawberry had **hermaphroditic** flowers. Such flowers possess both male and female parts. They are therefore capable of both pollinating other flowers and bearing fruit themselves. Hermaphroditic flowers are often referred to as **perfect flowers.**

Duchesne examined the flowers of the musky-flavored strawberry very carefully. They had both male and female parts, but there were two different forms of the flower. The plants that did not bear fruit had bigger flowers and

Antoine Nicholas Duchesne

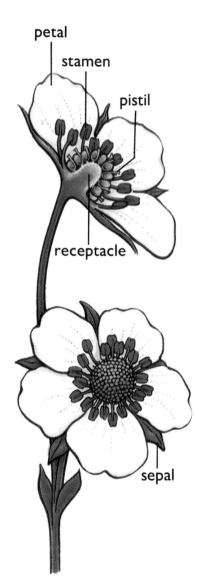

A perfect or hermaphroditic strawberry flower

robust stamens (male parts), but very small pistils (female parts). The fruit-bearing plants, on the other hand, had smaller flowers with normal pistils, but fewer stamens. What's more, the stamens didn't seem to contain any pollen.

Duchesne suspected that while the flowers looked as if they were hermaphroditic, each might actually function as only a male or only a female flower. If so, the musky-flavored strawberry would be **dioecious.** Dioecious species produce male and female flowers on separate plants.

In some of the first experiments ever performed with plants, Duchesne proved that the musky-flavored strawberry was indeed dioecious. Soon afterward, he began to work with the Chilean strawberry. After examining a specimen closely, he guessed that it too might be dioecious. That would explain the plant's fruiting problem. Captain Frézier would have originally collected the most healthy, productive plants he could find in Chile. Since male plants bear no fruit, Frézier probably brought back all female plants! New plants produced from the original plants would also be female. These female plants would not produce fruit if there were no male plants nearby to pollinate them.

Duchesne had learned that pollen from the common wood strawberry would not fertilize the female musky-flavored strawberry flower. And no one had ever seen fruit growing on Chilean strawberries planted near wood strawberries, either. Because he recognized some similarities between the Chilean strawberry and the musky-flavored strawberry, Duchesne reasoned that the pollen of one might be able to fertilize the pistils of the other. So he hand-pollinated the flowers of a Chilean strawberry with pollen from the musky-flavored strawberry. The result was the magnificent fruit that he presented to the king.

Louis XV responded to this unique gift by authorizing Duchesne to undertake a detailed study of the strawberry plant and to collect all the varieties of strawberries known in Europe for the royal garden. The young man took his job very seriously. Two years later, at age 19, he had not only gathered a unique collection of plants, but he had also written *Histoire Naturelle de Fraisiers,* a classic volume in natural history and one of the most thorough treatises ever written on the strawberry.

The Trianon forms part of Versailles, the palace of French royalty.

Duchesne's Experiments on the Musky-Flavored Strawberry

Duchesne's teacher, the famous botanist Bernard de Jussieu, had shown him certain peach trees that bore only male or only female flowers. The boy became suspicious that even though many types of strawberries had perfect flowers, the musky-flavored strawberry might produce flowers that were unisexual. He carried out these experiments to test his hypothesis:

Experiment A: Duchesne took a fertile (fruit-bearing) plant from a garden where it was growing among sterile (non-fruit-bearing) plants and put it in a pot, which he set by itself on the windowsill in his room. On this plant there were flowers in various stages of growth: open flowers, flowers in the process of developing into fruits, 5 or 6 withered blossoms, and about 20 buds. He cut off all the open flowers, which might already have been fertilized, got rid of all but three of the developing fruits, and marked several unopened buds with thread before pruning off the rest. The marked buds opened into flowers but withered without bearing fruit. But the infant fruits on the plant all developed into mature fruit. From this he learned that a known fertile plant would form no additional berries when isolated from other members of its species. The flowers on the same plant seemed incapable of fertilizing themselves. Presumably, pollen from different kinds of flowers (male flowers) was needed for the plant to produce new fruit.

Experiment B: Duchesne placed a fertile musky-flavored strawberry on a windowsill and surrounded it with four wood strawberry plants and one Virginia strawberry plant. No fruits developed, despite the presence of other species with vigorous pollen-producing stamens. He determined not only that a fertile (female) *Fragaria moschata* cannot fertilize itself, but that two other strawberry species seemed incapable of fertilizing it.

It took geneticists more than 60 years to provide a more detailed explanation for Duchesne's observations. The wood strawberry, they discovered, has only 14 chromosomes. It will not fertilize or crossbreed with the Chilean strawberry or with the Virginia strawberry, which have 56 chromosomes in each of their cells. The musky-flavored strawberry, which has 42 chromosomes, will occasionally cross with the North American species and produce fruit, but the resulting plants are not very healthy. The Chilean strawberry and the Virginia strawberry, which have an identical number of chromosomes (56), cross readily and produce strong hybrid offspring. Natural hybrids of the two had never occurred because in their wild state they grow thousands of miles apart.

Hybridizing the Strawberry

All living organisms are made up of cells. The structure and function of both plant and animal cells are controlled by threadlike bodies known as **chromosomes.** Chromosomes are found in the cell's nucleus, which acts as the cell's command center. Chromosomes are made up of large, complex molecules called DNA (deoxyribonucleic acid), RNA (ribonucleic acid), and proteins. Chromosomes control cells by regulating the production of two kinds of proteins—structural proteins and enzymes.

Genes are smaller parts of chromosomes. Genes control certain traits in living beings, such as size, color, and resistance to disease. Chromosomes are sometimes referred to as the genetic material of the cell, because they contain genes.

Different organisms have different numbers of chromosomes. Humans have 46 chromosomes, while pea plants usually have 14. The chromosomes are arranged in pairs, so that a single trait of an organism is controlled by two genes, one located on each member of the chromosome pair. The two genes of a chromosome pair may be identical or they may be different. If they are different, they might work together to express a certain trait. Or one gene—known as the dominant gene—can override the expression of the other, known as the recessive gene.

In a single organism, all body cells, or somatic cells, have the exact same chromosome composition. When somatic cells reproduce, they grow large, make a copy of their chromosomes, then divide in half. The result is two smaller cells, each with the same genetic content.

Some cells, however, have a different number of chromosomes. These are the reproductive cells, or sex cells. Sex cells, such as eggs and sperm, have exactly half

A microscopic photograph shows human chromosomes. The fluorescent dots highlight certain genes.

the number of chromosomes as somatic cells. The function of sex cells is to produce a new organism by uniting with another sex cell. Because they each have half the number of chromosomes, when two sex cells unite, the resulting new cell has the correct number of chromosomes. Also, half of the genes and chromosomes of the new organism come from the female sex cell and half come from the male sex cell. Thus the offspring inherits traits equally from its two parents.

Two hundred years ago, in Duchesne's time, none of this was known. The modern science of genetics—the study of how characteristics are passed from parents to offspring—dates from the early 1900s. In the 1700s, naturalists were not yet able to predict what traits would show up in the offspring that resulted from the crossing (intentional fertilization of one plant with pollen from another) of two different species or varieties of plants. Scientists only knew that plants could reproduce in two different ways, vegetatively and sexually, and that the results of these modes of reproduction were very different.

In **vegetative reproduction,** new plants originate from the vegetative parts of a plant—that is, from the somatic cells in leaves, roots, or stems. Rooting a leaf in water or growing a plant from a stem cutting are examples of vegetative reproduction. Another example is a runner, a slender stem that grows from the mother plant and takes root along the ground. Daughter plants reproduced vegetatively are genetically identical to the mother plant. No mixing of chromosomes has occurred.

Sexual reproduction in plants involves the uniting of sex cells and the production of seeds. In sexual reproduction, genetic material in the pollen joins with the genetic material in the ovules of the ovary to produce seeds. New plants grow from the resulting seeds. Daughter plants produced from sexual reproduction have an equal number of chromosomes from each of their parents.

One way strawberry plants reproduce is by sending out runners, from which new plants grow.

Sexual reproduction takes place easily between closely related plants (members of
the same species). It can also take place,
though with more difficulty, between less
closely related plants (different species of the
same genus). This is known as **hybridization.**
In Duchesne's time, people generally believed
that when two species hybridized, or interbred, the offspring would always be sterile and incapable of repro-
ducing. For example, when a donkey and a horse mated,
the offspring, a mule, was sterile. Botanists knew that
natural hybrids between plants could sometimes occur,
but they did not try to interbreed plant species because
they thought that hybrid offspring would always be inferior to their parents and probably unable to reproduce.

Duchesne showed that excellent fruit could result
from hybridization: European farmers could produce
big-fruited strawberries by interplanting their fields of
Chilean strawberries with rows of other compatible
strawberry plants. Moreover, he discovered that many
hybrid offspring could in fact reproduce. (The science

*A. N. Duchesne's sketch of
the pineapple strawberry plant*

of plant breeding—the development of new and improved plants through repeated hybridization—depends on this fact.) Finally, Duchesne correctly predicted that some of the hybrid offspring of dioecious plants (which bear only male or female flowers) could produce perfect flowers.

The First Modern Strawberry

In 1766 a description of a new strawberry, *Fragaria X ananassa,* also known as the pineapple strawberry, was included in the *Gardener's Dictionary.* It was reported to be a sturdy, big-fruited, perfect-flowered plant. The fruit had the shape of a pyramid and a flavor and perfume reminiscent of pineapple. No one knew the parentage of the plant or exactly how it had been developed. Duchesne, with his keen powers of observation, correctly identified the new strawberry as a hybrid—the result of the fertilization of a female Chilean strawberry by the pollen of the North American Virginia strawberry.

The pineapple strawberry was the hybrid that fulfilled all of Duchesne's predictions. Almost all modern strawberries are derived from the pineapple strawberry. Since the berry first appeared in the mid-1700s, plant breeders have rehybridized it again and again. They collected and grew the seed and selected many varieties from its most promising offspring.

These breeding programs have resulted in a variety of modern large-fruited strawberries that can be grown around the world. They bear names like Keens' Seedling, Tioga, Ozark Beauty, Howard 17, Keens' Imperial, Royal Sovereign, Tennessee Beauty, Pocahontas, Totem, Earlidawn, Black Prince, and Chandler. They have various colors and shapes and thrive differently in different climates. These hybrid berries have improved size, flavor, transportability, and resistance to disease.

Most modern strawberries are hybrids. They come in a variety of shapes and sizes, and they have a variety of names, such as Midway and Earliglow.

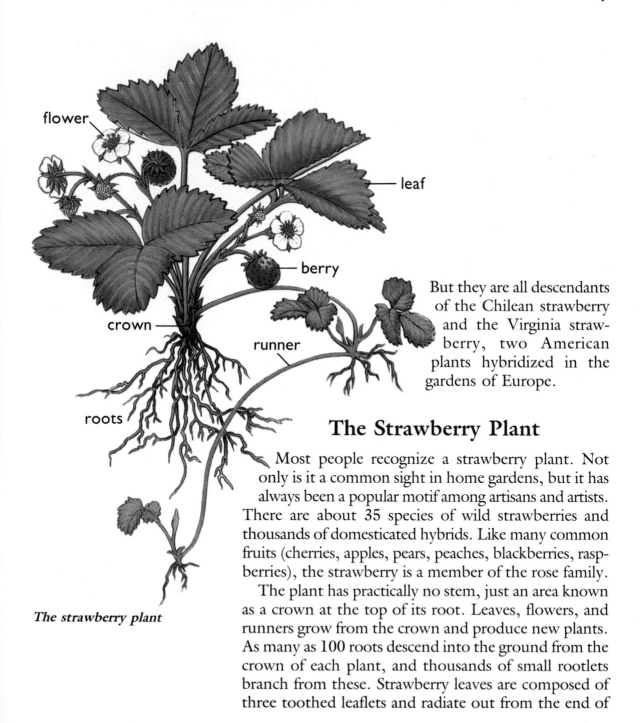

The strawberry plant

But they are all descendants of the Chilean strawberry and the Virginia strawberry, two American plants hybridized in the gardens of Europe.

The Strawberry Plant

Most people recognize a strawberry plant. Not only is it a common sight in home gardens, but it has always been a popular motif among artisans and artists. There are about 35 species of wild strawberries and thousands of domesticated hybrids. Like many common fruits (cherries, apples, pears, peaches, blackberries, raspberries), the strawberry is a member of the rose family.

The plant has practically no stem, just an area known as a crown at the top of its root. Leaves, flowers, and runners grow from the crown and produce new plants. As many as 100 roots descend into the ground from the crown of each plant, and thousands of small rootlets branch from these. Strawberry leaves are composed of three toothed leaflets and radiate out from the end of a long stem. The flowers, which grow in groups on leaf-less stalks, are white or pinkish. They have five round or oval petals arranged above green sepals, the small leaf-like structures that support and protect the petals. Below the sepals are more small leaflike structures called bracts.

Most strawberry flowers are perfect flowers, with both male and female parts. They usually have 20 to 35 short, golden, pollen-bearing stamens arranged around a cone-shaped structure called a **receptacle.** The pistils of the flowers are located on the receptacle. Strawberry flowers that are not perfect may be either male or female. Female flowers have normal pistils but stamens that do not develop normally. Male flowers have healthy stamens with lots of pollen, but they lack functional pistils.

Many kinds of flowers have several stamens but only one pistil. The pistil's ovary, however, often contains many ovules. The structure of the strawberry flower is different. Instead of having just one pistil, the strawberry has many pistils, each with a single ovary containing a single ovule.

Efficient pollination of all the pistils is very important for the development of strawberries. Both wind and bees carry pollen from one flower to another. After pollina-tion, ovules develop into seeds. Many fruits, such as tomatoes and peaches, are really matured ovaries, which become juicy and pulpy after fertilization. The red and juicy part of the strawberry is not the matured ovary, but the receptacle of the flower. Each ripe strawberry is really a large, mature receptacle.

The real fruits of the strawberry plant lie on the sur-face of the receptacle. Each receptacle contains hundreds of these hard, dried fruitlets, which botanists call **achenes.** Each achene contains a single seed. If you look at a strawberry under a magnifying glass, you will be able to see the fruitlets and dark, threadlike structures attached to some of them. These are remaining parts

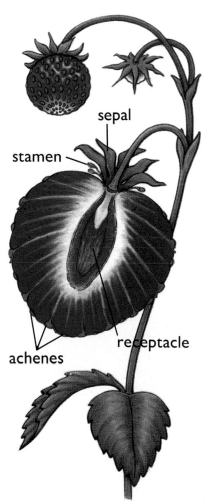

The strawberry fruit

of the pistils. Sometimes they float off when you put milk on strawberries.

The growth of the strawberry receptacle is controlled by the development of the fruitlets. The juicy red tissue will not develop around each pistil unless it is pollinated and produces the one-seeded fruit. For a big, symmetrical strawberry to develop, all of the pistils must be pollinated. If some are not, a small, misshapen, partially dried fruit results. You've probably seen some of these in the grocery store or on plants at home.

Soon after pollination, the flower petals start to wither. The green sepals and bracts, usually called the cap, remain as the receptacle grows. The cap has to be taken off when the strawberries are cleaned or hulled. If you look under the cap, you can see the withered stamens of the strawberry flower. It takes about a month from the time the flower opens to produce a strawberry that is ready to pick and eat.

Strawberry Flavor

Strawberry is one of the best-loved fruit flavors. Unfortunately, it is also one of the most difficult to capture or duplicate. The acid part of the taste comes from citric acid, which is also found in oranges and lemons. Strawberries also contain flavor compounds such as **esters, aldehydes,** and **ketones,** which are found in other fruits as well.

Examining Fresh Strawberry Flavor and Aroma

Put the same number of fresh and equally ripe strawberries in three bowls. Crush or mash the strawberries in bowl #1 with a fork and let them sit for five minutes uncovered. After five minutes, crush the strawberries in bowl #2 and let them sit for another five minutes uncovered. Finally, crush the strawberries in bowl #3. Immediately compare the aromas and flavors of the three bowls of strawberries. The freshly crushed berries in bowl #3 will be very fragrant and have a fully developed flavor. Those in bowl #2 will be less intense in both taste and smell, but still better than the fruit in bowl #1. A few minutes can make a lot of difference when serving strawberries!

The taste and aroma of fresh strawberries are very volatile. They deteriorate quickly when the berries are crushed, a fact that cooks know all too well. Flavor chemists have worked hard to find ways to capture and preserve the delicate strawberry flavor. One way is to freeze the whole berries for later use. Another technique involves freeze-drying—freezing the berries and removing most of the water from them at the same time.

A strawberry flavor extract can be made from the fresh or frozen fruits. The extract can then be preserved for long periods of time and added to other food products. To make strawberry flavor extract, the berries are mixed with alcohol, then crushed, pressed, and concentrated. Twenty or thirty pounds of fresh strawberries are necessary to produce about a gallon of extract.

Although strawberry flavor can be concentrated from the fruit, food scientists still do not know exactly what chemical substances give strawberries their unique taste and aroma. Flavors of fruits come from mixtures of chemicals that occur naturally in the fruit. In some fruits, like bananas, pears, and vanilla, a single compound can be identified as the one that provides the characteristic taste and smell. For others, such as apples and raspberries, the aroma and flavor are due to a blend of a small number of compounds. For still others, like apricots, peaches, and pineapples, the flavor and smell come from quite a large number of compounds. But strawberries present a unique problem.

When strawberries are analyzed in laboratories, more than 300 compounds can be identified in the fruit. But when scientists try to mix most of these compounds together in the right proportions, they cannot reproduce anything close to the true taste.

In the early 1900s, chemists discovered that certain aldehyde compounds, like aldehyde C_{16} (called strawberry aldehyde), produced an aroma that resembled that

Most strawberry-flavored products are artificially flavored, but the imitation flavor is nothing like the real thing.

of strawberry. Since then, other compounds with similar properties have been discovered. These compounds are used to provide or enhance strawberry flavor and smell in mass-produced foods, such as beverages, candy, ice cream, and chewing gum. Oddly enough, these artificial strawberry compounds do not occur in the strawberry plants themselves.

Synthetic strawberry flavorings don't come close to true strawberry flavor and aroma. If you put a few drops of imitation strawberry extract (available in most grocery stores) onto a sugar cube and compare it to the taste of juice pressed from a ripe strawberry, you will notice the difference immediately.

The Strawberry in Myth and Medicine

For centuries, strawberries have been associated with certain myths and superstitions. In Europe during the Middle Ages (A.D. 500–1500), pregnant women often did not eat the fruit for fear that their child would be born with a strawberry mark (a small, red-violet birthmark). Some people believed that when an infant died, he or she ascended to heaven in the form of a strawberry. Superstitious people often avoided eating the fruit, fearing it might be an act of cannibalism. Saint Hildegard von Bingen, a 12th-century nun and keen observer of plants, warned against eating strawberries, because they grew near the ground in what she referred to as "stale air."

In herbals (books describing the medicinal uses of plants) from as early as the mid-1400s, teas and potions made from all parts of the strawberry plant (roots, leaves, and fruit) were reported to be helpful for stomach pains, aching joints, and loose teeth, as well as for sore throats, bad breath,

Of Strawberries. Chap.370.

❋ The kindes.

THere be diuers forts of Strawberries, one red, another white, a thirde fort greene, and likewife a wilde Strawberrie, which is altogither barren of fruite.

1 *Fragaria & Fraga.*
Red Strawberries.

2 *Fragaria & Fraga fubalba.*
White Strawberries.

An illustration from an herbal published in 1597

fever, kidney stones, and bruises. The plant was also thought to help ease psychological problems such as depression. It is even said that in 1785 the great botanist Linnaeus cured himself of gout by following a diet of strawberries.

At least two medical properties of the plant have been scientifically confirmed in modern times. Nutritionists have shown that strawberries are exceptionally high in vitamin C, which some people claim is the best way to prevent the common cold. But in certain people, the berries are known to cause hives.

Growing Strawberries

Unlike cacoa trees and vanilla orchids, strawberries

are not difficult to grow. In fact, the berry is one of the most commonly grown fruits in the world. In the United States, strawberries are grown in every state. California produces more than half of the total domestic crop.

In commercial strawberry fields, the plants are often set out in long parallel rows of raised soil beds covered with plastic to keep the soil warm, the weeds down, and the moisture in. But you don't need a lot of land or fancy equipment to raise delicious berries—just rich soil, regular watering, and the right amount of sun.

If you want to plant strawberries in the ground, pick

The United States produces more strawberries than any other country. The average American eats about 3 pounds of strawberries a year.

a sunny spot, because strawberries do best in full sun. Avoid areas where other vegetables, particularly potatoes, tomatoes, eggplants, or peppers, have been grown in the past three years (these crops often carry soil-borne diseases that can destroy strawberry plants). Make sure the location is not infested with weeds. Before you plant, dig the soil well, and work in a lot of compost or manure so that the soil is rich and drains well. Add peat moss if you can, because peat makes the soil slightly acidic, which makes berries grow well.

Strawberries are usually planted in the spring, but they can also be planted in the fall. They are rarely planted from seed. Instead, most gardeners put out young plants grown from runners. Nurseries can advise you about the best types to plant in your area, and they usually carry inexpensive young plants. The young strawberry plants will be sold either in small pots or bundled together "bare-root," without soil around their root mass.

To grow big berries, you will have to remove all flowering stems and runners from your plants during the first growing season so that no fruit will develop. It's hard to do, but it will strengthen the mother plant and increase the number of runner plants and fruit the next year. If you don't want to do this, you can harvest the fruit the first year, but it will be a light crop of small berries.

Depending upon where you live and the variety of strawberries you are planting, the plant will flower anytime from late winter to midsummer and will bear fruit about a month later. The shape of your berries, again depending on the variety, may be round, conical, wedged, or angular, with or without a neck. Your berries may be as big as an egg or as small as a marble, depending on the health of your plants. The first berries your plant produces will always be the biggest, and later berries will become progressively smaller. To

How to Plant Strawberries

To plant bare-root strawberries, soak them in water or in a mixture of mud and water for a few hours. Cut off all dead or broken roots and all but one or two bright green leaves. Then dig a hole about six inches deep with a trowel, spread the roots apart with your fingers as you place the plant in the hole, and fill it with soil. Set the plant in the soil at the same depth that it grew in the nursery. If it is set too deep, the plant will be smothered and die. If it is planted too close to the surface, the plant will dry out. The crown or thick portion in the center of each plant should be set so that half of it is buried and half of it is above ground. Be sure to pack the soil around the roots of each plant. Dig your holes about 12 inches apart, and water the strawberries thoroughly after planting. It's a good idea to mulch the soil around the plants 3 or 4 inches deep with clean straw, pine needles, grass clippings, or even old newspapers. The mulch will keep moisture in the soil, discourage the growth of weeds, add nutrients to the soil, and keep the berries clean. The plants should be watered deeply at least once a week, or every three or four days if you live in a dry climate. If you live in an area where the winters are very cold, you should protect the plants once the temperature has dropped to about 20°F by covering the whole plant with a 4-inch blanket of mulch. Remove the mulch in early spring, when the plants start to grow again.

Make a hole with a trowel.

Rock trowel back and forth to pack sides of hole.

Place plant in hole, spread roots apart with fingers, and fill hole with soil.

— too deep
— just right
— too shallow

Growing Strawberries from Seed

It is not difficult to collect strawberry seeds. Pick some ripe strawberries. Fill a blender about half full of water and drop in 5 to 10 berries. Turn the blender on for about 20 seconds. Let the mixture stand for a few minutes, then pour off the pulp and most of the water. The good seeds will have sunk to the bottom. Scrape the seeds out of the blender, dry them on paper towels, and refrigerate them until you are ready to plant.

Unfortunately, since most strawberry plants grown today are hybrids, you cannot tell what kind of plants will grow from the hybrid seeds you have collected. Most of the offspring will not be very satisfactory plants.

If you want to grow strawberries from seed, try to buy seeds of Alpine strawberries at your local nursery or at the seed companies listed below. The Alpine strawberry is a runnerless selection of the European wood strawberry. It is not a hybrid and is usually grown from seed. The plant bears small but gourmet-quality fruit.

Sources of Alpine Strawberry seed:

Burpee Seed Co.
Warminster, PA 18991

Geo. W. Park Seed Co., Inc.
P.O. Box 31
Greenwood, SC 29647

prevent birds from getting to the berries before you do, try covering the bed with polyethylene netting or cheesecloth weighted down with bricks or rocks at the edges.

If you want to taste the strawberries at their very best, you should inspect your plants at least every other day in fruiting season and collect those berries that are perfectly red and ripe. Pick them with a short piece of stem attached. Do not leave picked berries in direct sunlight for more than 10 minutes, and keep them in shallow containers so they won't be bruised from being piled on top of one another. If you want to use the fruit for freezing or jam making, it is best to pick the berries slightly underripe, when they are a bit more firm. As long as the berries are whitish (not green), they will become red after picking, but they will never gain more flavor or sweetness.

If you are not going to eat the berries right away, keep them in a cool place, like a cellar or the bottom of a refrigerator. But if you want the best flavor, eat them within a few hours of picking them. Wash them just before they are to be used, and hull them (remove stem and cap) after they have been washed. Otherwise the berries will absorb water that will dilute their flavor. You

can eat the strawberries plain, sprinkled with a little sugar, topped with whipped cream, or prepared in one of the recipes at the end of this chapter.

If you don't have much space in your garden, you can grow strawberries in hanging baskets, old-fashioned clay strawberry jars, window boxes, or half barrels with slots cut into the sides. Alpine strawberries are particularly nice for planting in containers. They are a variety of the European wood strawberry and produce small but very delicious berries. The whole plant is delicate and does not produce runners like most other strawberries.

Propagating Strawberries

Strawberry patches need to be renewed every several years, but once you have one strawberry plant, you never need to buy another, because the plants reproduce themselves. They make new daughter plants very easily (this is called propagating). To renew your strawberry patch, allow the mother plants to make runners. Using a stone, clothespin, or even a handful of dirt, hold the runners in the place where you want them to grow. When the daughter plant is well rooted, you can cut the connection with the mother and get rid of the older plant. The young plant is now ready to grow and produce a good crop of strawberries.

Eating Strawberries

The best way to enjoy strawberries is to pop a freshly picked and washed ripe berry into your mouth. But some recipes have been included below to let you taste this exquisite flavor in different forms. Try them—you might agree with the great writer Jonathan Swift, who once said, "God could doubtless have made a better berry, but doubtless he never did."

If you grow your own strawberries, be sure to pick them when they are fully ripe and handle them gently.

RECIPE FOR: Strawberry Syrup
INGREDIENTS: 1 pound small strawberries
3/4 cup sugar
2 cups water

Wash and hull the strawberries. Combine
3/4 cup of sugar and 2 cups of water
in a saucepan and bring it to a boil.
When boiling hard, add the strawberries and remove the pan from
the stove. Put on a lid and
leave the berries to soak in
the sugar syrup until they are
cold. Blend the contents of

the pan in a blender or food processor, then pour the
liquid through a fine sieve or cheesecloth to remove the
seeds. You now have a strawberry syrup, which will
keep in the refrigerator for at least a week and can be
used in several ways.
    a) Add equal parts of strawberry syrup and vanilla
ice cream to a blender and blend for a couple of seconds.
The result is a luscious strawberry shake. Substitute
plain or sweetened yogurt for ice cream and you have
a real breakfast treat.
    b) Half fill a glass with strawberry syrup. Add
club soda or sparkling mineral water and ice for a
delicious and refreshing strawberry soda. Add a scoop

of ice cream and you have a strawberry ice cream
soda.
    c) Heat 2 cups of the syrup in a saucepan.
When it is hot, blend in 1/2 tablespoon of cornstarch
that has been dissolved in 1 tablespoon of water.
Bring the syrup to a boil, stirring it regularly, and
it will thicken and become a topping for pancakes
or ice cream.

RECIPE FOR: Old-Fashioned Strawberry Shortcake
INGREDIENTS: 2 cups flour
4 teaspoons baking powder
1/3 cup shortening, butter, or margarine
1/4 teaspoon salt
1 tablespoon sugar
1 egg, well beaten
2/3 cup milk
2 tablespoons sugar for berries
3 pints strawberries
1 cup heavy cream whipped
    with 2 tablespoons sugar

Preheat the oven to 400°F. Sift

together the flour, baking powder, salt, and one tablespoon sugar into a medium bowl. Cut in the vegetable
shortening using a pastry blender, two knives, or your
hands, until it is the consistency of cornmeal. Combine the egg and milk and stir into the flour mix
just until blended. Turn the dough out on a floured
surface and knead very lightly for 1 minute only. Pat
the dough into 1/2-inch thickness (don't roll it out thin!)
and cut it into rounds with a biscuit cutter or wide-
mouth glass. Arrange the biscuits on a lightly greased
baking pan and bake at 400°F for about 15 minutes,
or until puffed and golden. Cool slightly before using.
   While the biscuits are in the oven, wash, hull, and

cut the strawberries into thick pieces (or leave them
whole if they are small). Sprinkle the berries with 2
tablespoons sugar and let them stand for 10 minutes
or longer. When ready to serve, split the biscuits
and spoon berries between and on top of each biscuit, and top each with whipped cream.

SERVINGS: 8

RECIPE FOR: Strawberry Mousse
INGREDIENTS: 2 pints ripe juicy strawberries, or
20 ounces defrosted frozen strawberries
1 teaspoon lemon juice
2 packages unflavored gelatin
2/3 cup sugar
1/4 cup cold water
1 1/2 cups heavy cream
extra whipped cream for garnish

Combine strawberries (which
should be at room temperature),
sugar, and lemon juice in a
medium bowl and let them

steep for 30 minutes. Puree the mixture in a food
processor or blender and transfer it to a stainless
steel bowl. Sprinkle the gelatin over the cold water
in a small saucepan and let it stand for 5 minutes.
Then warm the mixture over low heat for a few seconds,
stirring constantly, until the gelatin dissolves completely
in the water. Stir the gelatin mixture into the puree.
   Place the stainless steel bowl containing the puree
into a larger bowl filled with ice and water. Every few
minutes, stir the puree, scraping the sides of the bowl
with a rubber spatula, until the puree starts to
thicken. This should take about 15 minutes. Whip the
cream until stiff peaks form, and gently fold it into

the puree. Spoon the mixture into a six-cup mold or
bowl, cover with plastic wrap, and chill for at least
3 hours before serving. This dessert is light, delicious,
and a lovely pink color.

SERVINGS: 6

Glossary

achenes (uh-KEENS)—small, single-seeded, hard, dry fruits.

aldehydes—a group of organic chemical compounds (compounds found in living things), usually derived from alcohols, that can easily be turned into acids.

amino acids—molecules that contain nitrogen and are the building blocks of proteins.

botanist—a scientist who studies plant life.

cacao—name of the tree that produces the beans from which chocolate is made. Also the name of the beans, or seedpods, of this tree.

chocolate liquor—the molten substance formed when the roasted nibs of the cacao beans are ground very fine. Chocolate liquor is the basic product used in all chocolate products (cocoa, molded chocolate candy, milk chocolate). When it cools and solidifies, it is often called baking chocolate.

chocolatl (chock-oh-LAH-tul)—Aztec drink made from fermented, dried, and ground cacao beans mixed with ground corn, water, vanilla, and spices. (Sometimes spelled *Xocolatl.*)

chromosomes—rod-shaped bodies found in the nucleus of cells. They carry genes, which determine the hereditary characteristics of organisms.

cocoa—chocolate product produced by removing most of the fat from chocolate liquor and then pulverizing the remaining block of hard, dry material.

cocoa butter—the vegetable fat contained in cacao beans. It is solid at room temperature but melts at body temperature. It is often used as a base for cosmetics and creams.

concentrate—to remove water from a substance so that it becomes stronger in flavor and reduced in quantity.

conching—the process developed by Rodolphe Lindt to knead chocolate after it has been refined, but before it is molded, to perfect the flavor and texture of the finished product.

cure—to preserve for later use. The curing of vanilla and cacao beans involves a controlled fermentation—in which chemical changes take place in the beans—followed by drying.

dioecious (dye-E-shus)—bearing male and female flowers on different plants.

distilling—the process of purifying and concentrating a substance by turning it into a vapor (gas) and then turning it back into a liquid.

domestication—the adaptation of a wild plant to the controlled environment of a garden or farm.

esters—a group of organic compounds that result from the chemical reaction between an acid and an alcohol. Animal and vegetable fats are esters.

ferment—to undergo a chemical change that occurs when organisms such as bacteria and yeasts convert sugars into alcohol and other products.

flavor precursors—chemical compounds that will contribute to the sensation of flavor after they have undergone some changes due to heat, fermentation, or processing.

gene—tiny part of a chromosome that determines the inheritance of certain characteristics.

hermaphroditic (her-MAF-roh-dih-tik)—possessing both male and female reproductive parts.

hybridization—the crossbreeding of two animals or plants of different species, or varieties.

ketones—a group of organic chemical compounds that has the following chemical grouping at one end: $C=O$. They are often important in determining the aroma of a compound.

nibs—roasted cacao beans with the seed coats removed. They are the raw material from which finished chocolate is ground.

ovules—the part of a flower that contains the female sex cell. After fertilization, ovules develop into seeds. Ovules are found in the ovary of a flower, which is part of the pistil.

perfect flowers—flowers with both male and female parts (hermaphroditic).

pistil—the female reproductive part of the flower, composed of an ovary, a style tube, and a sticky stigmatic surface that catches pollen.

pollen—yellowish powder (produced by stamens of flowers), which contains the male sex cells of the flower.

An early advertisement for Lindt chocolate

pollination—fertilization of the female part of a flower with pollen from the male part of a flower to produce seeds.

receptacle—the end of the flower stem, a sort of platform to which flower parts are attached. In the strawberry plant, after pollination of the flower, the receptacle forms the red, juicy part of the fruit.

receptors—nerve cells that receive sensory information or stimuli, such as taste and smell, and respond to this stimulation by sending nerve impulses to the brain.

refining—process in which steel rollers rotate over chocolate (a mixture of chocolate liquor, sugar, cocoa butter, and sometimes milk solids) to smooth the tiny particles of chocolate and sugar and eliminate any roughness or graininess in the final product.

runners—slender stems that take root along the ground and allow the parent plant to reproduce vegetatively.

sepals—the outer leaves that surround and protect the unopened bud of a flower. They often remain after the petals wither and the flower is developing into a fruit.

sexual reproduction—generation of new plants or animals through the union of sex cells—egg and sperm cells in animals, pollen and ovules in plants.

stamen—the male part of a flower, consisting of a slender stem called the filament, and the anther, which contains pollen.

tempering—process of gradually cooling chocolate.

theobromine (the-uh-BROH-meen)—a stimulant found in chocolate.

tlilxochitl (tlil-ZOH-chee-tul)—Aztec name for the dried, cured vanilla bean.

vanillin—chemical component in a vanilla pod responsible for much of its flavor and smell.

vegetative reproduction—nonsexual generation of offspring from a parent plant. The offspring are always identical to the parent.

A young man sells hot cocoa in Paris around 1700.

METRIC CONVERSION CHART

When you know:	multiply by:	to find:
Length		
inches	25.0	millimeters
inches	2.5	centimeters
feet	30.0	centimeters
yards	0.9	meters
miles	1.6	kilometers
Area		
square miles	2.6	square kilometers
acres	0.4	hectares
Mass (weight)		
ounces	28.0	grams
pounds	0.45	kilograms
Volume		
teaspoons	5.0	milliliters
tablespoons	15.0	milliliters
fluid ounces	30.0	milliliters
cups	0.24	liters
pints	0.47	liters
quarts	0.95	liters
gallons	3.8	liters
Temperature		
Fahrenheit	5/9 (*after* subtracting 32)	Celsius

A Note on Cooking

Whenever you cook, it's important to keep in mind some safety rules. Even experienced cooks must follow these rules in the kitchen.

1. Always wash your hands before handling food.
2. Thoroughly wash all raw vegetables and fruits to remove dirt, chemicals, and insecticides.
3. Use a cutting board when cutting up vegetables and fruits. Cut in a direction away from your fingers.
4. Long hair or loose clothing can catch fire near the burners of a stove. If you have long hair, tie it back before you start cooking.
5. Turn all pot handles away from you so that you will not catch your sleeves or jewelry on them.
6. Always use a pot holder to steady hot pots or to take pans out of the oven. Don't use a wet cloth on a hot pan, because the steam it produces can burn you.
7. If you get burned, hold the burn under cold running water.
8. If grease catches fire, throw baking soda or salt at the bottom of the flame. (Water will *not* put out a grease fire.) Call for help, and try to turn the burners off.

Acknowledgments

The photographs and illustrations in this book are reproduced through the courtesy of: pp. 2–3, 9, 16 (top), 25, 33 (top), 38, 39, 43, 57, 64 (bottom), 65, 67, 82, 85 (top), 94, Andy King; pp. 8, 56, Cadbury Limited; pp. 10, 11, 12, 21, 22, 24, 32, 47, 49, 68, 80, 88, 89, 98, Laura Westlund; pp. 13, 14 (top), 72, 93, W. Atlee Burpee & Co.; pp. 14 (bottom), 34, 62, 63 (top), Jim Simondet; p. 15, Puerto Rico Federal Affairs Administration; p. 16 (bottom), American Museum of Natural History; pp. 17, 29 (bottom), 58 (left), Independent Picture Service; pp. 18, 42, Library of Congress; pp. 19, 23, 26, 28, McCormick & Company, Inc.; pp. 27, 29 (top), 30, 35, Kathy Krezek-Larson/Frontier Cooperative Herbs; p. 33 (bottom), Burch Communications Inc.; pp. 40, 41 (bottom), BBC Hulton Picture Library; p. 41 (top), Bruckman, F. E., Relatio brevis historica-botanica medica de Avellana Mexicana, vulgo cacao dicta, Brunswick, 1728; pp. 44 (left), 46, 48, 52 (top), 55 (top), 59, 60, 107, Lindt & Sprüngli (USA) Inc.; pp. 44 (right), 50, 52 (bottom), 54, 55 (bottom), 58 (right), 61 (inset), Chocolate Manufacturers Association; p. 53, Eliot Elisofon, Eliot Elisofon Archives, Museum of African Art, Smithsonian Institution; pp. 61, 63 (bottom), Hershey Foods Corporation; p. 64 (top), Nestlé; p. 66, Kathy Raskob/IPS; p. 69, Robert L. and Diane Wolfe; pp. 73, 91, © Valorie Hodgson: PHOTO/NATS, INC.; p. 75 (top), Herbarius Latinus Moguntiae (Latin Herbal of Mainz) of 1484; p. 76 (left), © Jean Baxter: PHOTO/NATS, INC.; p. 76 (right), © Jo-Ann Ordano: PHOTO/NATS, INC.; p. 77, Révue Horticole (1879); p. 79 (top), The Bancroft Library; p. 79 (bottom), The Gardener's Chronicle, vol. 54, July 1913; pp. 81, 108, Bettmann; p. 83, Laboratory of David Ward, Yale University, from Science Magazine, Volume 247, January 5, 1990; pp. 84, 86, 87, 98, 101, Nourse Farms, Inc.; p. 85 (bottom), Bibliothèque du Muséum National d'Histoire Naturelle, Paris; pp. 90, 92, 97, California Strawberry Advisory Board; p. 95, Herball by John Gerard (1597); p. 96, © Deborah M. Crowell: PHOTO/NATS, INC.; p. 102, © Priscilla Connell: PHOTO/NATS, INC.

BAKER'S is a registered trademark of Kraft General Foods, Inc. Photograph on page 45 reproduced with permission. We would like to acknowledge with thanks the invaluable contribution Lindt & Sprüngli has made to this publication.

Maps on pages 31, 51, and 74 are by Laura Westlund.